Non-linear and Non-stationary Time Series Analysis

To Ruth, Michael, and Marie

Non-linear and Non-stationary Time Series Analysis

M. B. Priestley

Department of Mathematics, University of Manchester Institute of Science and Technology

1988

ACADEMIC PRESS

Harcourt Brace Jovanovich, Publishers

London San Diego New York
Berkeley Boston Sydney Tokyo Toronto

ACADEMIC PRESS LIMITED
24/28 Oval Road, London NW1 7DX

United States Edition Published by
ACADEMIC PRESS INC.
San Diego, CA 92101

British Library Cataloguing in Publication Data

Priestly, M. B.
Non-linear and non-stationary time series analysis.
1. Time series. Analysis
I. Title
519.5′5

ISBN 0-12-564910-X

Printed in Great Britain by
St Edmundsbury Press Ltd, Bury St Edmunds, Suffolk

Preface

The past ten years have witnessed some major developments in the field of time series analysis. The twin assumptions of linearity and stationarity which underlie so much of conventional time series analysis have finally been abandoned, and the subject has moved in a new direction. This new direction is, of course, the study of non-linear models, and it is this area, together with the treatment of spectral analysis of non-stationary processes, which forms the substance of this book. Non-linear models offer both great excitement and great challenges. The mathematical ideas involved are much more complex than those of linear models, and the statistical problems of model identification and parameter estimation are similarly more intricate. However, once these nettles are grasped, the richness of the new models is quite fascinating. Experience so far gained has shown that even simple non-linear models can capture types of time series behaviour which it would be impossible to describe with linear models.

So far most of the work on non-linear models is scattered throughout the literature, and the main purpose of this book is to bring these ideas together and to present them within a common unified framework. The scope of the book is roughly as follows. Chapter 1 presents a summary of the basic concepts of stationary processes and their spectral analysis. Chapter 2 gives a review of the classical theory of linear models, including a discussion of state-space representations. Chapter 3 deals with general forms of non-linear models, and introduces the topics of Volterra series expansions, polyspectra, and tests for non-linearity. Chapter 4 describes some special (finite parameter) classes of non-linear models, with special emphasis on the treatment of bilinear, threshold autoregressive, and exponential autoregressive models. Chapter 5 extends the discussion of non-linear models to the general class of "state-dependent" models which is shown to include, as special cases, the specific models treated in Chapter 4. Chapter 6 moves to the study of non-stationary processes, and presents the theory of evolutionary (time-dependent) spectra, together with a description of some practical applications of this theory. Finally, Chapter 7 presents an extension of the classical Kolmogorov–Wiener theory of prediction and filtering to the case of non-stationary processes.

Although the book contains a brief review of basic time series concepts (Chapter 1), readers are assumed to have a working knowledge of conventional time series analysis. Those who lack sufficient background knowledge may find it useful to consult the author's previous book, *Spectral Analysis and Time Series* (Academic Press, London) for additional information on relevant topics.

I must express my sincere thanks to my colleague Dr T. Subba Rao, and former colleagues Dr Valerie Haggan (now Valerie Ozaki) and Professor H. Tong. Much of the work described in Chapters 4 and 5 was developed whilst we were all colleagues in the Department of Mathematics at the University of Manchester Institute of Science and Technology (UMIST), and I benefited greatly from the many stimulating discussions in which we were constantly engaged. I am indebted also to a number of postgraduate students at UMIST who made significant contributions to the initial development of special types of non-linear models. In particular, Dr S. M. Heravi played an important role in the construction of the complex computer programmes required for the numerical study of the state-dependent models described in Chapter 5.

The typing of the manuscript has been undertaken by various persons over the past few years; to them and to all who have contributed to this work I express my gratitude.

Finally, I am grateful to the Royal Statistical Society, the Applied Probability Trust, TIETO Limited (publishers of the *Journal of Time Series Analysis*), and Springer-Verlag for permission to reproduce various figures and graphs which have appeared in their publications.

M. B. PRIESTLEY

Contents

Chapter 1

Introduction and Background Theory

1.1 INTRODUCTION

A *time series* is a record of the values of any fluctuating quantity measured at different points of time. For example, we may have a record of daily temperatures, a record of the voltage in an electrical circuit measured at intervals of 1 sec, a record of the monthly price index of a commodity, or an EEG (electroencephalograph) record measuring the electrical activity between two points in a person's brain. The first three records are examples of time series recorded at a *discrete* set of time points (every day, every second, every month), whereas the fourth is an example of a time series recorded *continuously* over a period of time. Both types of series arise in practical applications, but the "discrete" type occurs more frequently than the "continuous" type, and in this book we shall be concerned mainly with the former type of series.

One of the most important features of the vast majority of time series is that the values recorded at different time points are all influenced, at least in part, by some *random mechanism.* A record of daily temperatures, for example, will exhibit irregular patterns of variation which cannot be adequately described by a simple mathematical formula; an EEG record will show a similar behaviour. In the case of voltage measurements made on a simple electrical network we may argue that by appealing to the laws of electromagnetic theory we could, in principle, derive an explicit mathematical formula for the voltage as a function of time. However, it is important to remember that the recorded voltages will inevitably contain measurement errors, and these errors are, by their nature, random quantities. Thus, even in this case, the recorded voltages would be influenced by random elements.

Let us now use the symbol X_t to denote the value of the series under study at the time instant t. In the language of probability theory we would say that, for each t, X_t is a *random variable*, with an associated *probability distribution*, $p_t(x)$. If, as we shall assume, X_t has, for each t, a continuous range of possible values, then $p_t(x)$ would correspond to a *probability density function*, so that, for example, the mean, μ_t, and variance, σ_t^2, of X_t would be then given by

$$\mu_t = \int_{-\infty}^{\infty} x p_t(x)\, dx, \tag{1.1.1}$$

$$\sigma_t^2 = \int_{-\infty}^{\infty} \{x - \mu_t\}^2 p_t(x)\, dx. \tag{1.1.2}$$

Note, however, that $p_t(x)$ describes the probability distribution of the time series only at one particular time instant. To describe the full probabilistic properties of the complete series we need to specify the *joint probability distribution* of the values of the series at all the time instants under consideration. Thus, if we are interested in studying the behaviour of the series at times $t = 1, 2, \ldots, N$, then we need to consider the (multivariate) joint probability distribution of the N random variables, $X_1, X_2, \ldots, X_N$.

1.2 STATISTICAL ANALYSIS OF TIME SERIES

One of the characteristic features which distinguishes time series data from other types of statistical data is the fact that, in general, the values of the series at different time instants will be *correlated*. (One would expect today's temperature to be correlated to some extent with the temperatures recorded on previous days.) In more general terms this means that the random variable X_t will be correlated with the random variables $X_{t-1}, X_{t-2}, X_{t-3}, \ldots,$ and with $X_{t+1}, X_{t+2}, X_{t+3}, \ldots$. A basic problem in analysing a time series is to study the pattern of the correlation between values at different time instants, and to try to construct statistical models which "explain" the correlation structure of the series.

During the past 50 years or so time series analysis has become a highly developed subject, and there are now well-established methods for fitting a wide range of models to time series data—as described, for example, in the books by Anderson (1971), Box and Jenkins (1970), Brillinger (1975), Chatfield (1975), Hannan (1970), Jenkins and Watts (1968), Koopmans (1975), and Priestley (1981). However, virtually all the established methods rest on two fundamental assumptions; namely that (i) the series is *stationary* (or can be reduced to stationarity by some simple transformation, such as differencing), and (ii) the series conforms to a *linear* model. Assumption (i) means, in effect, that the main statistical properties of the series remain constant over time, and (ii) means that the values of the observed series can be represented as linear combinations of present and past values of a "strictly random" (or "independent") series. Needless to say, both these assumptions are mathematical idealizations which, in some cases, may be valid only as approximations to the real situation. In practical applications the most one could hope for is that, for example, over the observed time interval the series would not depart "too far" from stationarity for the results of the analysis to be invalid. It would seem natural, therefore, to try to extend the methods of classical time series analysis to deal with the more general situations of non-stationarity and non-linearity, and indeed over the last few years there has been a strong interest in these extensions of basic time series methodology. Experiences gained so far indicate that non-linear and non-stationary time series analysis is certainly feasible from a practical point of view, and, in many cases, can lead to improved methods of model fitting and forecasting. Most of the recent work in this area has appeared only in journals and conference proceedings, and the purpose of this book is to bring together the various developments which have taken place, and to present them within a unified theoretical framework.

To facilitate further discussion of non-linear and non-stationary phenomena, we summarize below some of the basic and well-known concepts of the theory of stationary time series. This summary is, of necessity, rather brief and is intended mainly for readers who are already acquainted with this theory. Readers who are not familiar with the subject may find it peferable to read a more detailed account, such as, for example, that given in Chapters 3 and 4 of the author's book *Spectral Analysis and Time Series* (Academic Press, 1981). Henceforth, references to this book will be indicated by the abbreviation *Spec. Anal.*

1.3 BASIC THEORY OF STATIONARY TIME SERIES

When we denote a time series by X_t we almost invariably think of the symbol t as representing real time. Occasionally, however, t may represent some other physical quantity—such as length. (For example, we may be interested in studying variations in the thickness of a length of yarn at different points along its length.) We therefore refer to t more generally as the "parameter" of the series, and call the series *discrete parameter* or *continuous parameter* according to whether t takes a discrete or a continuous set of values. In the former case we would usually assume that the values of t were taken at equally spaced intervals, so that typically t would take the discrete set of values $0, \pm 1, \pm 2, \ldots$. (It is convenient, from a mathematical point of view, to allow t to take both positive and negative values.) To avoid possible confusion between the two cases we adopt the convention throughout the book of denoting a discrete parameter series by X_t and a continuous parameter series by $X(t)$. As previously remarked, most of the discussion will, in fact, deal with discrete parameter series, but in general corresponding results hold also for the continuous parameter case.

Stationarity

A series $\{X_t\}$ is called "stationary" if, loosely speaking, its statistical properties do not change with time. More precisely, $\{X_t\}$ is said to be *completely stationary* if, for any set of times $t_1, t_2, \ldots, t_n$, and any integer k, the joint probability distribution of $\{X_{t_1}, X_{t_2}, \ldots, X_{t_n}\}$ is identical with the joint probability distribution of $\{X_{t_1+k}, X_{t_2+k}, \ldots, X_{t_n+k}\}$. Less stringently, we say that $\{X_t\}$ is *weakly stationary* (or *stationary up to order 2*), if only the joint moments up to order 2 of the above probability distributions exist and are identical. A series which is weakly stationary is usually referred to simply as *stationary*, and in this case we have

$$E[X_t] = \mu, \quad \text{say,}$$

$$E[X_t^2] = \mu_2', \quad \text{say,}$$

where μ, μ_2' are constants independent of t, so that

$$\text{var}[X_t] = \sigma^2, \quad \text{say, a constant independent of } t.$$

Moreover, $E[X_tX_s]$ will be a function only of $(t-s)$, and hence $\text{cov}\{X_t, X_s\}$ will similarly be a function only of $(t-s)$. Thus, for a stationary series the mean and variance of X_t remain constant over time, and the covariance between any two values, X_t, X_s, depends only on the separation between the time points and not on their individual locations.

Autocovariance and autocorrelation functions

If we consider $\text{cov}\{X_t, X_{t+r}\}$ we see that this quantity depends only on r and not on t. We may thus write,

$$\begin{aligned} \text{cov}\{X_t, X_{t+r}\} &= E[(X_t - \mu)(X_{t+r} - \mu)] \\ &= R(r), \quad \text{say} \qquad r = 0, \pm 1, \pm 2, \ldots. \end{aligned} \tag{1.3.1}$$

The function $R(r)$ is called the *autocovariance function* of $\{X_t\}$. If we now write, for each r,

$$\rho(r) = \frac{R(r)}{R(0)} \tag{1.3.2}$$

then $\rho(r)$ is called the *autocorrelation function.* Noting that $R(0) \equiv \sigma^2$ it follows that $\rho(r)$ is the correlation coefficient between X_t and X_{t+r}.

Gaussian processes

$\{X_t\}$ is called a *Gaussian process* if, for all $t_1, t_2, \ldots, t_n$, the set of random variables $\{X_{t_1}, X_{t_2}, \ldots, X_{t_n}\}$ has a multivariate normal distribution. Since a multivariate normal distribution is completely specified by its means and variance–covariance matrix, it follows that, in this case, if $\{X_t\}$ is stationary to order 2 it must be completely stationary.

Spectral density functions

When $R(r)$ decays to zero sufficiently fast so that $\sum_{r=-\infty}^{\infty} |R(r)| < \infty$, we may introduce the (discrete) Fourier transform of $R(r)$, namely

$$h(\omega) = \frac{1}{2\pi} \sum_{r=-\infty}^{\infty} R(r)\, e^{-i\omega r}, \qquad -\pi \leq \omega \leq \pi. \tag{1.3.3}$$

The function $h(\omega)$ is called the *(power) spectral density function* of X_t (or simply the *spectrum* of X_t) and inverting (1.3.3) we obtain

$$R(r) = \int_{-\pi}^{\pi} e^{i\omega r} h(\omega)\, d\omega, \qquad r = 0, \pm 1, \pm 2, \ldots. \tag{1.3.4}$$

It may be shown that even if $R(r)$ does not decay fast enough for the function $h(\omega)$ to exist, we may still represent $R(r)$ in a form similar to (1.3.4), using a generalized Fourier transform, namely

$$R(r) = \int_{-\pi}^{\pi} e^{i\omega r}\, dH(\omega), \tag{1.3.5}$$

where $H(\omega)$ is a non-decreasing function with $H(-\pi) = 0$, $H(\pi) = \sigma^2$, called the *integrated spectrum* or *spectral distribution function.* The representation (1.3.5) is embodied in a result known as the "Wiener–Khintchine theorem" (see *Spec. Anal.*, Chapter 4). Inverting (1.3.5) gives

$$H(\omega) = \sigma^2 \left(\frac{\omega + \pi}{2\pi} \right) + \frac{1}{2\pi} \left[\sum_{r=-\infty}^{-1} + \sum_{r=1}^{\infty} \right] \frac{e^{-ir\omega}}{-ir} R(r). \tag{1.3.6}$$

Of course, when $h(\omega)$ exists we have $dH(\omega) = h(\omega)\, d\omega$, so that

$$H(\omega) = \int_{-\pi}^{\omega} h(\theta)\, d\theta.$$

Corresponding to (1.3.5) there exists an analogous representation for X_t, which, when $E[X_t]=0$, takes the form

$$X_t = \int_{-\pi}^{\pi} e^{i\omega t}\, dZ(\omega), \qquad t=0, \pm 1, \pm 2, \ldots, \tag{1.3.7}$$

where $Z(\omega)$ is a (complex-valued) stochastic process with orthogonal increments, i.e. $E[dZ(\omega)\, dZ^*(\omega')]=0$, $\omega \neq \omega'$, (* denoting complex conjugate), and is related to $H(\omega)$ by

$$E[|dZ(\omega)|^2] = dH(\omega). \tag{1.3.8}$$

Equation (1.3.6) is a fundamental result known as the "spectral representation" of X_t (see *Spec. Anal.*, Chapter 4), and it tells us, in effect, that any discrete parameter stationary series can be represented as a "sum" of sines and cosines involving a continuous range of frequencies over $(-\pi, \pi)$, and with random amplitudes, $|dZ(\omega)|$, and random phases, $\arg\{dZ(\omega)\}$. Equation (1.3.8) provides the physical interpretation of $dH(\omega)$; roughly speaking, $dH(\omega)$ is the mean-square amplitude of the component in X_t with frequency ω. More precisely, when X_t represents some physical process it may be shown that $H(\pi)$ $(=\sigma^2)$ is a measure of the average *total power* dissipated by the process, and $dH(\omega)$ $(=h(\omega)\, d\omega)$ represents the contribution to the total power from the components in X_t with frequencies between ω, $\omega + d\omega$. The function $h(\omega)$ thus represents the distribution of the power *density* over frequency.

Continuous parameter series

For a continuous parameter process, $\{X(t)\}$, the autocovariance and autocorrelation functions are defined as in (1.3.1), (1.3.2), but now r takes a continuous range of values over $(-\infty, \infty)$. The power spectral density function (when it exists) is now given by

$$h(\omega) = \frac{1}{2\pi} \int_{-\infty}^{\infty} R(r)\, e^{-i\omega r}\, dr \tag{1.3.9}$$

and correspondingly, when $R(r)$ is continuous,

$$R(r) = \int_{-\infty}^{\infty} e^{i\omega r} h(\omega)\, d\omega. \tag{1.3.10}$$

The more general representation of $R(r)$ is given by

$$R(r) = \int_{-\infty}^{\infty} e^{i\omega r}\, dH(\omega) \tag{1.3.11}$$

where $H(\omega)$ is a non-decreasing function over $(-\infty, \infty)$, with $H(-\infty)=0$, $H(\infty)=\sigma^2$.

When $X(t)$ is "stochastically continuous" with zero mean, it has the spectral representation

$$X(t) = \int_{-\infty}^{\infty} e^{it\omega}\, dZ(\omega) \tag{1.3.12}$$

with $\{dZ(\omega)\}$ an orthogonal process over $(-\infty, \infty)$.

Complex-valued series

So far we have assumed that X_t is a real-valued series. However, the theory is easily extended to cover the more general case where X_t is allowed to take complex values, i.e. where $X_t = U_t + iV_t$, say, U_t, V_t being real-valued series. In this case, the definitions of the mean and autocovariance function are modified to

$$\mu = E[X_t] = E[U_t] + iE[V_t],$$
$$R(r) = E[\{X_t - \mu\}^*\{X_{t+r} - \mu\}] \tag{1.3.13}$$

(where again * denotes complex conjugate). The variance is now given by

$$\sigma^2 = R(0) = E[|X_t - \mu|^2].$$

With this modification to $R(r)$, the definitions previously given for $h(\omega)$ and $H(\omega)$ remain valid for the complex-valued case, as does the spectral representation of X_t.

Multivariate series

In some situations we may be presented with data from several series recorded simultaneously over time. Suppose, for example, that we have p (discrete parameter) series, $X_{1,t}, X_{2,t}, \ldots, X_{p,t}$, $t = 0, \pm 1, \pm 2, \ldots$. If we assume that each series is stationary, we may define its respective autocovariance, autocorrelation, and spectral density function as above, and the study of these functions will then provide information about the *individual* structure of each of the series. However, if we study the properties of each series separately we may well miss important information regarding the "cross-links" between the different series. To investigate possible relationships between the series we have to introduce new functions called the "cross-covariance" and "cross-correlation" functions, defined as follows. In addition to each individual series being stationary, we now assume further that the collection of p series are *jointly stationary*, i.e. that for each i, j, $\text{cov}\{X_{i,t}, X_{j,s}\}$ is a function of $(s-t)$ only. The function

$$R_{ij}(r) = \text{cov}\{X_{i,t}, X_{j,t+r}\}, \qquad r = 0, \pm 1, \pm 2, \ldots \tag{1.3.14}$$

is then called (for $i \neq j$) the *cross-covariance* function between $X_{i,t}$ and $X_{j,t}$ (with $X_{i,t}$ "leading" $X_{j,t}$). The *cross-correlation* function is correspondingly defined by

$$\rho_{ij}(r) = R_{ij}(r)/\{R_{ii}(0)R_{jj}(0)\}^{1/2}. \tag{1.3.15}$$

(When $i=j$, $R_{ii}(r)$ is, of course, the autocovariance function of $X_{i,t}$ and similarly for $R_{jj}(r)$.) For each r, the auto- and cross-covariances may be formed into a *covariance matrix*, $\boldsymbol{R}(r)=\{R_{ij}(r)\}$, $i=1,\ldots,p$, $j=1,\ldots,p$ i.e. $\boldsymbol{R}(r)$ has $R_{ij}(r)$ in the ith row and jth column.

Assuming that for each i, $\sum_{r=-\infty}^{\infty}|R_{ii}(r)|<\infty$, the spectral density functions of each of the individual series may be evaluated as in (1.3.3) as the Fourier transforms of the $R_{ii}(r)$, i.e. we may write

$$h_{ii}(\omega) = \frac{1}{2\pi}\sum_{r=-\infty}^{\infty} R_{ii}(r)\, e^{-i\omega r}, \qquad i=1,\ldots,p. \tag{1.3.16}$$

In addition, if $\sum_{r=-\infty}^{\infty}|R_{ij}(r)|<\infty$, (all i,j), we may introduce also the Fourier transforms of the $\{R_{ij}(r)\}$, namely

$$h_{ij}(\omega) = \frac{1}{2\pi}\sum_{r=-\infty}^{\infty} R_{ij}(r)\, e^{-i\omega r}, \qquad i,j=1,\ldots,p \tag{1.3.17}$$

which is then called the *cross-spectral density function* between $X_{i,t}$ and $X_{j,t}$. The *spectral matrix* is then defined by $\boldsymbol{h}(\omega)=\{h_{ij}(\omega)\}$, $i,j=1,\ldots,p$.

The covariance and spectral matrices may be defined in a more concise form by introducing the column vector, $\boldsymbol{X}_t=\{X_{1,t}, X_{2,t},\ldots,X_{p,t}\}'$, in which case $\boldsymbol{R}(r)$ may be written (assuming $E[\boldsymbol{X}_t]=0$)

$$R(r) = E[\boldsymbol{X}_{t+r}\boldsymbol{X}_t^*], \tag{1.3.18}$$

(where * here denotes both conjugation and transposition), and we now have

$$\boldsymbol{h}(\omega) = \frac{1}{2\pi}\sum_{r=-\infty}^{\infty} \boldsymbol{R}(r)\, e^{-i\omega r}, \tag{1.3.19}$$

the inversion of which gives

$$\boldsymbol{R}(r) = \int_{-\pi}^{\pi} e^{i\omega r}\boldsymbol{h}(\omega)\, d\omega. \tag{1.3.20}$$

It is easily shown that the joint stationarity of the p series implies that $\boldsymbol{R}^*(r)=\boldsymbol{R}(-r)$ (each r), and hence that $\boldsymbol{h}(\omega)$ is a *Hermitian matrix*, i.e. that for each ω, $\boldsymbol{h}^*(\omega)=\boldsymbol{h}(\omega)$.

Using the spectral representations of each of the series, namely (cf. (1.3.7))

$$X_{i,t} = \int_{-\pi}^{\pi} e^{i\omega t}\, dZ_i(\omega) \tag{1.3.21}$$

the joint stationarity property implies further that the $\{dZ_i(\omega)\}$ are *orthogonal* and *cross-orthogonal*, i.e.

$$E[dZ_i^*(\omega)\, dZ_j(\omega')] = 0, \qquad \omega \neq \omega', \text{ all } i, j.$$

Substituting (1.3.21) into (1.3.18), using the above orthogonality properties and comparing the resulting expression with (1.3.20) now gives us an alternative expression for the cross-spectral density function, $h_{ij}(\omega)$, namely

$$h_{ij}(\omega)\, d\omega = E[dZ_j^*(\omega)\, dZ_i(\omega)]. \tag{1.3.22}$$

Writing $\boldsymbol{dZ}(\omega) = \{dZ_1(\omega), dZ_2(\omega), \ldots, dZ_p(\omega)\}'$, the spectral matrix can thus be expressed in the form

$$\boldsymbol{h}(\omega)\, d\omega = E[\boldsymbol{dZ}(\omega)\, \boldsymbol{dZ}^*(\omega)], \tag{1.3.23}$$

i.e. $\boldsymbol{h}(\omega)\, d\omega$ may be interpreted as the variance–covariance matrix of the vector $\{\boldsymbol{dZ}(\omega)\}$.

1.4 LINEAR REPRESENTATIONS FOR STATIONARY PROCESSES

We have seen that a (univariate) stationary process may be expressed in terms of its spectral representation, (1.3.7), as a "sum" of sine and cosine functions with orthogonal (i.e. uncorrelated) "coefficients", $\{dZ(\omega)\}$. However, there is an alternative linear representation of stationary processes which also involves orthogonal variables, and which may be derived as follows.

Consider a zero mean stationary process, X_t, with spectral representation

$$X_t = \int_{-\pi}^{\pi} e^{it\omega}\, dZ(\omega). \tag{1.4.1}$$

Suppose that X_t has an "absolutely continuous spectrum", i.e. that its spectral density function $h_X(\omega)$ exists for all ω, so that from (1.3.8) we have

$$E[|dZ(\omega)|^2] = h_X(\omega)\, d\omega. \tag{1.4.2}$$

Since the LHS of the above equation is clearly non-negative, we have $h_X(\omega) \geq 0$, all ω, and consequently we can find some (possibly complex-valued) function, $\phi(\omega)$, such that we may write,

$$h_X(\omega) = |\phi(\omega)|^2 = \phi(\omega)\phi^*(\omega). \tag{1.4.3}$$

Also, setting $r = 0$ in (1.3.4), we obtain,

$$\int_{-\pi}^{\pi} h_X(\omega)\, d\omega = R(0) = \text{var}\{X_t\}, \tag{1.4.4}$$

so that (assuming X_t has finite variance), $\phi(\omega)$ is quadratically integrable, and hence can be expanded in a Fourier series,

$$\phi(\omega) = \sum_{u=-\infty}^{\infty} \gamma_u e^{-i\omega u}, \tag{1.4.5}$$

say. Assuming now that $h_X(\omega)$ is strictly positive over the range $(-\pi, \pi)$, so that $|\phi(\omega)| > 0$, all ω, we set $dz(\omega) = dZ(\omega)/\phi(\omega)$, and rewrite (1.4.1) as

$$X_t = \int_{-\pi}^{\pi} e^{i\omega t} \phi(\omega)\, dz(\omega).$$

Substituting (1.4.5) into the above (and changing the order of summation and integration) gives

$$X_t = \sum_{u=-\infty}^{\infty} \gamma_u \left\{ \int_{-\pi}^{\pi} e^{i(t-u)\omega}\, dz(\omega) \right\},$$

or

$$X_t = \sum_{u=-\infty}^{\infty} \gamma_u \varepsilon_{t-u}, \tag{1.4.6}$$

where

$$\varepsilon_t = \int_{-\pi}^{\pi} e^{it\omega}\, dz(\omega).$$

We may note that the RHS of (1.4.6) exists as a mean-square limit since we have

$$2\pi \sum_{u=-\infty}^{\infty} \gamma_u^2 = \int_{-\pi}^{\pi} |\phi(\omega)|^2\, d\omega = \int_{-\pi}^{\pi} h(\omega)\, d\omega = \operatorname{var}\{X_t\} < \infty.$$

Now ε_t is a sequence of *uncorrelated* variables since the spectral density function of ε_t is given by

$$h_\varepsilon(\omega)\, d\omega = E[|dz(\omega)|^2] = E[|dZ(\omega)|^2]/|\phi(\omega)|^2 - \frac{h_X(\omega)\, d\omega}{|\phi(\omega)|^2} = d\omega.$$

Thus, $h_\varepsilon(\omega)$ takes the constant value unity for all ω, and the inversion formula (1.3.4) then shows that the autocovariance function of ε_t vanishes everywhere except at the origin. (The process ε_t is usually called a *purely random* or *white noise* process.)

To summarize, we have shown that any stationary process with a continuous spectrum may be represented in the form (1.4.6) as a linear combination of past, present, and future values of an uncorrelated process, ε_t. If we impose a stronger condition on $h_X(\omega)$, we can obtain a more specialized result in which X_t is represented as a linear combination of past and present values only of an uncorrelated process, i.e. we can derive a representation of the form (1.4.6) but with the summation over extending only from 0 to ∞ rather than from $-\infty$ to ∞. The extra condition required is that $\log h_X(\omega)$ be integrable, i.e.

$$\int_{-\pi}^{\pi} \log\{h_X(\omega)\}\, d\omega > -\infty. \tag{1.4.7}$$

Under this condition we can find a function $\phi(\omega)$ satisfying (1.4.3) and such that it has a "one-sided" Fourier series

$$\phi(\omega) = \sum_{u=0}^{\infty} g_u e^{-i\omega u}, \tag{1.4.8}$$

(see *Spec. Anal.*, Chapter 10, p. 733). Proceeding as above we now obtain the "one-sided" linear representation for X_t

$$X_t = \sum_{u=0}^{\infty} g_u \varepsilon_{t-u}. \tag{1.4.9}$$

Equation (1.4.9) is an important result and plays a basic role in the theory of stationary time series. (It has particularly important applications in the theory of "linear prediction", and in this context it forms a special case of a more general result known as the *Wold decomposition*, which provides a canonical representation for stationary series whose spectra may contain both discrete and continuous components—see *Spec. Anal.*, Chapter 10, Section 10.1.5).

Linear filters

A useful way of interpreting (1.4.9) is to think of the uncorrelated process ε_t as the "input" to a system, and X_t as the corresponding "output". More generally, consider a system with input U_t and output V_t—as shown in Fig. 1.1.

If the system is linear and time invariant, the relationship between V_t and U_t can be expressed in the form

$$V_t = \sum_{u=0}^{\infty} a_u U_{t-u}, \tag{1.4.10}$$

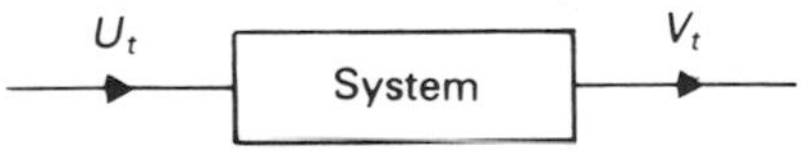

Fig. 1.1. Linear system.

where $\{a_u\}$ is a sequence of constant called the *impulse response function* of the system. Now suppose that U_t is a zero mean stationary series with spectral density function $h_U(\omega)$. If $h_U(\omega)$ is a bounded function of ω, and $\sum_{u=0}^{\infty} a_u^2 < \infty$, then it may be shown that V_t is also a zero mean stationary series and has spectral density function $h_V(\omega)$ given by

$$h_V(\omega) = h_U(\omega)|A(e^{-i\omega})|^2, \tag{1.4.11}$$

where the function $A(z)$ is defined by

$$A(z) = \sum_{u=0}^{\infty} a_u z^u \tag{1.4.12}$$

(see *Spec. Anal.*, Chapter 4, Section 4.12). A linear system of the above form is usually referred to as a *filter*, and the function $A(z)$ is then called the *transfer function* of the filter.

If we now replace U_t, V_t, by ε_t, X_t, respectively, then (1.4.9) provides a description of X_t as the output of a linear filter with input ε_t. Since ε_t is an uncorrelated process, its spectral density function, $h_\varepsilon(\omega)$, is a constant; in fact, we have $h_\varepsilon(\omega) = \sigma_\varepsilon^2/2\pi$, where σ_ε^2 denotes the variance of ε_t. Applying the general result (1.4.11) to (1.4.9) then gives

$$h_X(\omega) = \frac{\sigma_\varepsilon^2}{2\pi}|G(e^{-i\omega})|^2, \tag{1.4.13}$$

with

$$G(z) = \sum_{u=0}^{\infty} g_u z^u. \tag{1.4.14}$$

Moreover, $G(e^{-i\omega}) \equiv \phi(\omega)$ (where $\phi(\omega)$ is given by (1.4.8)), and hence, setting $\sigma_\varepsilon^2 = 2\pi$, we see that (1.4.13) is, in fact, identical with the factorization (1.4.3), which was the starting point from which we derived the linear representation (1.4.9).

Chapter 2

Review of Linear Models

2.1 MODEL BUILDING IN TIME SERIES ANALYSIS

The very simplest type of time series which may arise is that generated by a *strict white noise* process, and we denote this special type of series by $\{e_t\}$. Such a series consists merely of a sequence of *independent* random variables, and, if stationary, its mean μ and variance σ^2 are the same at all time points. The correlation between values at different time points is, of course, zero, and we may therefore write,

$$E\{e_t\} = \mu, \qquad \text{var}\{e_t\} = \sigma^2, \quad \text{all } t,$$

$$\text{cov}\{e_t, e_s\} = 0, \quad \text{all } t \neq s.$$

However, a general time series, $\{X_t\}$, will have a much more complicated structure. The *mean function* of $\{X_t\}$, namely $\mu_t = E\{X_t\}$, will in general change over time due, for example, to the presence of trends and seasonal components. More importantly, the variation about the mean function, as described by the series $X'_t = \{X_t - \mu_t\}$, will be *autocorrelated*, i.e. there will, in general, be correlation between the value of X'_t at different time points. In studying a general time series of this form we try to construct a *model* which describes both the nature of the mean function and the dependence or correlation between values at different time points. Let us suppose, for simplicity, that we have removed all trends and seasonal components, so that we may assume, in effect, that the mean function of $\{X_t\}$ is zero for all t. Our task now is to "explain" the autocorrelation properties of the series, and we try to do this by finding some relationship between past, present, and future values which reduces the series to *strict white noise.*

Suppose that the series is observed only at a discrete set of time points, say, $t = 0, \pm 1, \pm 2, \ldots$. When we speak of a *model* for such a series we mean a relationship between $\ldots X_{t-2}, X_{t-1}, X_t, X_{t+1}, X_{t+2}, \ldots$ which produces a zero mean strict white noise process, e_t. Thus, a model for $\{X_t\}$ would, in its most general form, be described by an equation of the form

$$h\{\ldots X_{t-2}, X_{t-1}, X_t, X_{t+1}, X_{t+2}, \ldots\} = e_t, \tag{2.1.1}$$

where $h\{\cdot\}$ is some prescribed function. An analogy with simple regression models might be helpful; suppose we are given a number of observations $\{x_i, y_i\}$, on two variables x, y (with the x values predetermined) and we wish to find a functional relationship between x and y. We will have determined the "best" relationship between x and y when we have found that function $f(\cdot)$ such that $y_i = f(x_i) + e_i$, where $\{e_i\}$ is a sequence of independent zero mean random variables, i.e. when we have reduced $\{y_i - f(x_i)\}$ to strict white noise. (Otherwise, if the residuals from the regression model were not strict white noise, there would be further structure in the relationship between x and y which we had not "explained" by the model.)

The most general problem of time series model building may now be stated in the following terms:

General time series analysis: given some observations on $\{X_t\}$, find that function $h\{\cdot\}$ which reduces the LHS of (2.1.1) to a strict white noise process.

If we can find such a function $h\{\cdot\}$ (which, moreover, is such that the model (2.1.1) is "invertible", i.e. is such that (2.1.1) may be "solved" so as to express each X_t as a function of $\ldots e_{t-2}, e_{t-1}, e_t, e_{t+1}, e_{t+2}, \ldots$) then we have found the best possible model for the time series $\{X_t\}$.

However, posed in this very general form, with $h\{\cdot\}$ completely arbitrary, the problem is of course an impossible one to solve. For, given only a finite amount of data on $\{X_t\}$ we could not possibly hope to determine a specific function $h\{\cdot\}$ from among the class of all possible functions. One way of surmounting this difficulty is to reduce the scale of the problem drastically by considering only the class of *linear models*. This class of models is the one on which virtually the whole of standard time series is based, and corresponds to the case where in (2.1.1) $h\{\cdot\}$ is restricted to being a *linear* function of $\ldots X_{t-2}, X_{t-1}, X_t, X_{t+1}, X_{t+2}, \ldots$. In this case (2.1.1) can be written as

$$\sum_{u=-\infty}^{\infty} h_u X_{t-u} = e_t, \tag{2.1.2}$$

where $\{h_u\}$ is a sequence of constants. Equation (2.1.2) gives the most general form of linear model, but it is "anticipative" in the sense that X_t is allowed to depend (at least in part) on future as well as past values of the series. In practice, we would usually assume (for physical reasons) that X_t depends only on past values, i.e. we would set $h_u = 0$, $u < 0$, and rewrite (2.1.2) in the "one-sided" form,

$$\sum_{u=0}^{\infty} h_u X_{t-u} = e_t. \tag{2.1.3}$$

Introducing the backward shift operator, B, defined by $BX_t = X_{t-1}$, $B^2 X_t = X_{t-2}$, etc., (2.1.3) can now be expressed in the form,

$$H(B)X_t = e_t \tag{2.1.4}$$

where

$$H(z) = \sum_{u=0}^{\infty} h_u z^u. \tag{2.1.5}$$

Equation (2.1.4) may be "solved" to express X_t as a linear function of present and past values of e_t. Thus, inverting (2.1.4) formally, we may write,

$$X_t = H^{-1}(B)e_t \tag{2.1.6}$$

and if $H^{-1}(z)$ is analytic for $|z| \leq 1$ (i.e. if $H(z)$ has no zeros inside and on the unit circle) we may write further

$$H^{-1}(z) = g_0 + g_1 z + g_2 z^2 + \cdots. \tag{2.1.7}$$

We can now write X_t explicitly as

$$X_t = \sum_{u=0}^{\infty} g_u e_{t-u} \tag{2.1.8}$$

or

$$X_t = G(B)e_t, \tag{2.1.9}$$

where

$$G(z) = \sum_{u=0}^{\infty} g_u z^u. \tag{2.1.10}$$

The above condition on the zeros of $H(z)$ ensures that the RHS of (2.1.8) converges in mean-square—see, for example, *Spec. Anal.*, Chapter 3. We have, of course, the obvious relationship between $H(z)$ and $G(z)$, namely $G(z) \equiv H^{-1}(z)$, or $H(z)G(z) \equiv 1$. Equation (2.1.8) provides an alternative formulation for the general non-anticipative linear model in which X_t is expressed as a (in general infinite) linear combination of present and past values of a strict white noise process, $\{e_t\}$.

It is often useful to interpret (2.1.8) within the context of "linear systems theory", in which case we would think of $\{e_t\}$ as the *input process* and $\{X_t\}$ as the *output process.* With this interpretation the function $G(z)$ (or more conventionally, its form on the unit circle, namely $G(e^{-i\omega})$), is called the *transfer function* of the model—as discussed in Section 1.4. Note that the linear model (2.1.8) (or (2.1.3)) is completely determined by the form of $G(z)$.

The problem of general linear time series model fitting thus takes the form:

> *General linear time series analysis: given some observations on $\{X_t\}$, estimate the sequence $\{h_u\}$ (or equivalently, the sequence $\{g_u\}$) which reduces the LHS of (2.1.3) to strict white noise.*

This poses yet another impossible problem since clearly given only a finite amount of data on $\{X_t\}$ we cannot hope to estimate an infinite number of arbitrary parameters. We must therefore reduce the scale of the problem still further, and the standard approach is to assume that the function $H(z)$ (or $G(z)$) takes a known mathematical form involving only a finite number of unknown parameters. This approach leads us into the area of *finite parameter linear models,* and particular types of finite parameter linear models can be characterized in terms of the special forms of $G(z)$ to which they correspond. We now consider three particularly important classes of finite parameter models.

2.2 AR, MA, AND ARMA MODELS

In the context of practical time series model fitting we use a standard set of finite parameter models which arise as special cases of the general linear model (2.1.8). Thus, if we assume that $G^{-1}(z)$ $(\equiv H(z))$ may be approximated by a finite-order polynomial, say,

$$G^{-1}(z) = 1 + a_1 z + \cdots + a_k z^k \tag{2.2.1}$$

(where we have assumed further that $a_0 = 1$), then (2.1.8) reduces to the AR(k) (*autoregressive model of order* k) form

$$X_t + a_1 X_{t-1} + \cdots + a_k X_{t-k} = e_t. \tag{2.2.2}$$

On the other hand, if we approximate to $G(z)$ by a finite-order polynomial, say,

$$G(z) = 1 + b_1 z + \cdots + b_l z^l \tag{2.2.3}$$

then (2.1.8) reduces to the MA(l) (*moving average of order l*) form,

$$X_t = e_t + b_1 e_{t-1} + \cdots + b_l e_{t-l}. \tag{2.2.4}$$

(Again we have assumed, without loss of generality, that $b_0 = 1$.)

If $G(z)$ and $G^{-1}(z)$ are "well-behaved" functions we can, to an arbitrary degree of accuracy, approximate either by a finite polynomial of sufficiently high degree. It follows, therefore, that we can, in general, approximate a general linear model by either an AR or MA model of sufficiently high order. However, we can obtain a more "parsimonious" finite parameter representation of $G(z)$ by approximating it by a *rational function.* If we take $G(z)$ to have the form,

$$G(z) = \frac{1 + b_1 z + \cdots + b_l z^l}{1 + a_1 z + \cdots + a_k z^k} = \frac{\beta(z)}{\alpha(z)}, \quad \text{say}, \tag{2.2.5}$$

then (2.1.8) reduces to the ARMA(k, l) (*mixed autoregressive/moving average of order* (k, l)) model,

$$X_t + a_1 X_{t-1} + \cdots + a_k X_{t-k} = e_t + b_1 e_{t-1} + \cdots + b_l e_{t-l}. \tag{2.2.6}$$

This model can be written in operator form as,

$$\alpha(B) X_t = \beta(B) e_t. \tag{2.2.7}$$

For the above model to possess a stationary solution of the form (2.1.8) we require that the corresponding form of $G(z)$ (namely (2.2.5)) be analytic for $|z| \leq 1$, i.e. *that $\alpha(z)$ has no zeros inside or on the unit circle.* We assume also that in (2.2.5) $G(z)$ has been expressed in "reduced form", i.e. that the polynomials $\alpha(z)$, $\beta(z)$ contain no common factors, and $a_k \neq 0$, $b_l \neq 0$. The model fitting problem now reduces to the following:

> *Finite parameter linear time series analysis: given some observations of $\{X_t\}$, estimate the unknown parameters k, l, $a_1, \ldots, a_k$, $b_1, \ldots, b_l$.*

Under this formulation the problem becomes tractable, and there are now several well-established methods for fitting ARMA models of the form (2.2.6) to time series data (see, for example, *Spec. Anal.*, Chapter 5). In fact, the ARMA model (and its variants, such as the ARIMA and "seasonal" ARMA models) form the basis of virtually the whole of "classical" time series model fitting methods.

The slightly more general ARIMA models which have been studied, in particular by Box and Jenkins (1970), correspond to the case where the operator $\alpha(B)$ in (2.2.7) contains a factor of the form $(1-B)^d$. In this case $G(z)$ contains a dth-order pole at $z = 1$ and consequently $\{X_t\}$ is no longer a stationary process—although its dth difference, $\Delta^d X_t$, would, in general, be stationary.

Autocovariance function and spectral density function of ARMA model

Assuming that (2.2.7) generates a stationary process, its autocovariance function, $R(r)$, may be shown to satisfy the difference equation,

$$R(m)+a_1R(m-1)+\cdots+a_kR(m-k)=0, \qquad m\geq \max(k, l+1), \tag{2.2.8}$$

(see *Spec. Anal.*, Chapter 3). The solution of (2.2.8) will, in general, contain a mixture of decaying exponential terms and damped oscillatory terms. The process $\{X_t\}$ then has a *purely continuous spectrum*, the corresponding spectral density function, $h_X(\omega)$, being the Fourier transform of $R(r)$. However, $h_X(\omega)$ is most conveniently evaluated using the theory of linear filters described in Section 1.4. Using the result (1.4.13), we have,

$$h_X(\omega)=\frac{\sigma_e^2}{2\pi}|G(e^{-i\omega})|^2=\frac{\sigma_e^2}{2\pi}\frac{|\beta(e^{-i\omega})|^2}{|\alpha(e^{-i\omega})|^2}. \tag{2.2.9}$$

Thus, for a stationary ARIMA model, the *spectral density function is a rational function of* $(e^{-i\omega})$.

2.3 STATE SPACE REPRESENTATIONS

There is an interesting way of rewriting the general ARMA model (2.2.6) using the so-called *state space* (or *Markovian*) representation which provides a very compact description of any finite parameter linear model. The basic idea rests simply on the well-known result that any finite-order linear differential or difference equation can be expressed as a vector first-order equation. For example, if we take the AR(2) model,

$$X_t+a_1X_{t-1}+a_2X_{t-2}=e_t, \tag{2.3.1}$$

and write

$$x_t^{(2)}=X_t, \qquad x_t^{(1)}=-a_2X_{t-1} \qquad (=-a_2x_{t-1}^{(2)})$$

then (2.3.1) may be rewritten as,

$$\begin{bmatrix} x_t^{(1)} \\ x_t^{(2)} \end{bmatrix}=\begin{bmatrix} 0 & -a_2 \\ 1 & -a_1 \end{bmatrix}\begin{bmatrix} x_{t-1}^{(1)} \\ x_{t-1}^{(2)} \end{bmatrix}+\begin{bmatrix} 0 \\ 1 \end{bmatrix}e_t. \tag{2.3.2}$$

To recover X_t from the vector $[x_t^{(1)}, x_t^{(2)}]'$, we now write,

$$X_t=(0, 1)\begin{bmatrix} x_t^{(1)} \\ x_t^{(2)} \end{bmatrix} \tag{2.3.3}$$

The pair of equations (2.3.2), (2.3.3) are completely equivalent to (2.3.1), but whereas (2.3.1) involves a two-stage dependence (so that X_t is non-Markovian), (2.3.2) involves only a one-stage dependence so that $[x_t^{(1)}, x_t^{(2)}]'$

is a vector Markov process. (The device of expressing (2.3.1) in the form (2.3.2) may alternatively be viewed as a special case of the technique of writing certain types of non-Markov processes involving only finite stage dependence as vector Markov processes.)

Exactly the same approach may be used to write the general ARMA model in the state space form. Thus, writing $n = \max(k, l+1)$, we first rewrite (2.2.6) as

$$X_t + a_1 X_{t-1} + \cdots + a_n X_{t-n} = e_t + b_1 e_{t-1} + \cdots + b_{n-1} e_{t-n+1} \qquad (2.3.4)$$

(where $a_i = 0$, $i > k$, $b_j = 0$, $j > l$), and it is then easy to verify that (2.3.4) can be written in state space form as,

$$\boldsymbol{x}_{t+1} = \boldsymbol{F}\boldsymbol{x}_t + \boldsymbol{G}e_t, \qquad (2.3.5)$$

$$X_t = \boldsymbol{H}\boldsymbol{x}_t, \qquad (2.3.6)$$

where the $n \times n$ matrix $\boldsymbol{F}$ is given by,

$$\boldsymbol{F} = \begin{bmatrix} 0 & 0 & \dots & 0 & -a_n \\ 1 & 0 & \dots & 0 & -a_{n-1} \\ 0 & 1 & \dots & 0 & -a_{n-2} \\ \dots & \dots & \dots & \dots & \dots \\ 0 & 0 & \dots & 1 & -a_1 \end{bmatrix}, \qquad (2.3.7)$$

the $n \times 1$ matrix $\boldsymbol{G}$ by,

$$\boldsymbol{G}' = [b_{n-1}, b_{n-2}, \dots, b_1, 1], \qquad (2.3.8)$$

the $1 \times n$ matrix $\boldsymbol{H}$ by,

$$\boldsymbol{H} = [0, 0, \dots, 0, 1], \qquad (2.3.9)$$

and the $n \times 1$ vector $\boldsymbol{x}_t$ is partitioned as,

$$\boldsymbol{x}_t = [x_t^{(1)}, x_t^{(2)}, \dots, x_t^{(n)}]'.$$

(In fact, $x_t^{(n)}$ is simply the variable X_t in (2.3.4) and $x_t^{(n-1)}, x_t^{(n-2)}, \dots,$ are defined recursively in terms of $x_t^{(n)}$.) In this formulation, $\boldsymbol{x}_t$ is called the *state vector,* $\boldsymbol{F}$ is called the *system matrix,* $\boldsymbol{G}$ the *input matrix,* and $\boldsymbol{H}$ the *observation matrix.* As previously noted, the condition for (2.3.4) to represent a stationary process is that $\alpha(z)$ has no zeros inside or on the unit circle—in which case the matrix $\boldsymbol{F}$ is "*stable*", i.e. all its eigenvalues have modulus less than 1, and consequently the process $\boldsymbol{x}_t$ is stationary (*Spec. Anal.*, Chapters 3, 11).

It should be emphasized that the specific expressions for $\boldsymbol{F}$, $\boldsymbol{G}$, and $\boldsymbol{H}$ given above provide merely one way of writing (2.3.4) in state space form. For example, we may write down an alternative representation for the AR(2) model (2.3.1), by using the state vector, $x_t^{(1)} = X_{t-1}$, $x_t^{(2)} = X_t$, leading to,

$$\begin{bmatrix} x_t^{(1)} \\ x_t^{(2)} \end{bmatrix} = \begin{bmatrix} 0 & 1 \\ -a_2 & -a_1 \end{bmatrix} \begin{bmatrix} x_{t-1}^{(1)} \\ x_{t-1}^{(2)} \end{bmatrix} + \begin{bmatrix} 0 \\ 1 \end{bmatrix} e_t \tag{2.3.10}$$

$$X_t = (0, 1) \begin{bmatrix} x_t^{(1)} \\ x_t^{(2)} \end{bmatrix}. \tag{2.3.11}$$

The corresponding generalization for the ARMA model (2.3.4) then takes the form (2.3.5), (2.3.6), where now $\boldsymbol{F}$, $\boldsymbol{G}$, $\boldsymbol{H}$, are given by

$$\boldsymbol{F} = \begin{bmatrix} 0 & 1 & 0 & \cdots & 0 \\ 0 & 0 & 1 & \cdots & 0 \\ 0 & 0 & 0 & \cdots & 1 \\ -a_n & -a_{n-1} & -a_{n-2} & \cdots & -a_1 \end{bmatrix} \tag{2.3.12}$$

$$\boldsymbol{G} = [0, 0, \ldots, 1]', \tag{2.3.13}$$

$$\boldsymbol{H} = [b_{n-1}, b_{n-2}, \ldots, b_1, 1]'. \tag{2.3.14}$$

The expressions (2.3.7)–(2.3.9) and (2.3.12)–(2.3.14) constitute two well-known *canonical forms* for the state space representation of linear models, the latter being termed the *phase-variable canonical form*.

These two alternative forms highlight the fact that there is no unique state space representation of a linear model, and it is important to note that, for given $\alpha(z)$, $\beta(z)$, there is, in general, a multitude of different forms of $\boldsymbol{x}_t$, $\boldsymbol{F}$, $\boldsymbol{G}$, and $\boldsymbol{H}$, in terms of which (2.2.6) can be written in the form (2.3.5), (2.3.6). Even if we fix n (the dimension of $\boldsymbol{x}_t$), we may still obtain an equivalent state space representation by making a "similarity transformation", $\boldsymbol{x}_t^* = \boldsymbol{T}\boldsymbol{x}_t$, ($\boldsymbol{T}$ being an arbitrary, non-singular $n \times n$ matrix), and setting $\boldsymbol{F}^* = \boldsymbol{T}\boldsymbol{F}\boldsymbol{T}^{-1}$, $\boldsymbol{G}^* = \boldsymbol{T}\boldsymbol{G}$, $\boldsymbol{H}^* = \boldsymbol{H}\boldsymbol{T}^{-1}$. However, a more significant point is that a given form of (2.2.6) may be rewritten in the form (2.3.5), (2.3.6), using state vectors of different dimensions, i.e. the relationship between (2.2.6) and (2.3.5), (2.3.6), does not entail a unique value of n—unless we impose certain conditions on $\boldsymbol{F}$, $\boldsymbol{G}$, $\boldsymbol{H}$, which ensure, in effect, that $\boldsymbol{x}_t$ does not contain "redundant" components. Equations (2.3.5), (2.3.6) are said to provide a *minimal (or irreducible) realization* when based on a state vector $\boldsymbol{x}_t$ with the lowest possible dimension, and the dimension of $\boldsymbol{x}_t$ may then be called the *dimension of the model* (2.2.6).

The notion of "minimal realizations" is closely related to the concepts of "controllability" and "observability" introduced by Kalman (1963). If, for the moment, we think of e_t as a physical input (rather than an unobservable strict white noise process), with X_t as the corresponding output and the state vector $\boldsymbol{x}_t$ having dimension n, then the system described by (2.3.5), (2.3.6), is said to be *completely controllable* if the matrix

$$[\boldsymbol{G}\colon \boldsymbol{F}\,\boldsymbol{G}\colon \ldots\colon \boldsymbol{F}^{n-1}\,\boldsymbol{G}] \tag{2.3.15}$$

is of full rank n, while the system is said to be *completely observable* if the matrix

$$[\boldsymbol{H}'\colon (\boldsymbol{H}\boldsymbol{F})'\colon \ldots\ (\boldsymbol{H}\,\boldsymbol{F}^{n-1})']' \tag{2.3.16}$$

is of full rank n. ("Controllability" means, in effect, that starting from an initial state $\boldsymbol{x}^{(0)}$ at time t_0, there exists an input which "steers" the system into any desired state $\boldsymbol{x}^{(1)}$ at some subsequent time t_1, and (2.3.15) is a well-known condition for a linear system to have this property. On the other hand, when the "observability" condition (2.3.16) holds, it is possible, in principle, to determine the state vector $\boldsymbol{x}_t$ explicitly from a sequence of observations on $\{X_t\}$ and $\{e_t\}$; see, for example, Rosenbrock (1970)). Kalman (1963) proved a basic result in linear systems theory which states that *a realization is minimal if and only if the corresponding system (represented by the triplet of matrices,* **F**, **G**, **H**) *is completely controllable and completely observable.* Kalman has pointed out that the "input/output" relationship between $\{e_t\}$ and $\{X_t\}$ determines only that part of the system which is completely controllable and completely observable; thus minimal realizations may be regarded as the standard (i.e. non-pathological) mode of representing input/output relationships.

Akaike's representation

Akaike (1974a, 1974b) has given a very general treatment of finite-order linear models, and has shown, using a geometrical approach, that a minimal realization can be derived by selecting the state vector $\boldsymbol{x}_t$ as any basis of the "predictor space at time t", $R(t+|t-)$, namely the space spanned by the linear least squares predictors of $X_t, X_{t+1}, X_{t+2}, \ldots$, given $e_t, e_{t-1}, e_{t-2}, \ldots$. The crucial feature of this space is that, for models described by finite-order difference equations of the form (2.2.6), it has *finite dimension*, $n = \max(k, l+1)$. A minimal realization can now be constructed by choosing the state vector (at time t) as *any* basis of $R(t+|t-)$. If we again think of the model as a physical system with e_t as the "input" and X_t as the "output", the above specification of the state vector is in complete accord with the physical interpretation of the "state" of the system at time t_0 as that set of quantities which, together with the input for all time points $\geq t_0$, uniquely determines the output for all time points $\geq t_0$. (When, for example, the relationship between input and output is described by a finite-order difference or differential equation, the "state" at time t_0 is simply the set of "initial conditions" required to determine the solution of the difference or differential equation for all $t \geq t_0$.) Intuitively, therefore, we may think of the state at any time point as the totality of information on the future output contained within the past input, and this is precisely the information contained within the above set of predictors, or equivalently, within any set of basis variables of the space spanned by these predictors.

2.4 LINEAR MODELS AND STATIONARITY

Before considering non-linear models it is natural to enquire as to the degree of generality of linear models as descriptions of stationary processes. In other words, if we assume that X_t is (weakly) stationary, can we always represent it by a general linear model of the form (2.1.8)? The results of Section 1.4 show that a linear representation which *superficially* resembles (2.1.8) can always be constructed for any stationary process X_t provided only that it has a *purely continuous spectrum*, with spectral density function $h_X(\omega)$, say. In this case we saw that we could write X_t in the form (cf. (1.4.6)),

$$X_t = \sum_{u=-\infty}^{\infty} \gamma_u \varepsilon_{t-u}, \tag{2.4.1}$$

and moreover, if $h_X(\omega)$ satisfies the condition (1.4.7), we could replace (2.4.1) by the "one-sided" form,

$$X_t = \sum_{u=0}^{\infty} \gamma_u \varepsilon_{t-u}. \tag{2.4.2}$$

At first sight (2.4.2) looks very much like (2.1.10), and we appear to have shown that any stationary process with a purely continuous spectrum satisfying the condition (1.4.7) can be described by a general linear model. However, this is untrue since we have overlooked the crucial point that the process e_t in (2.1.10) has to be a sequence of *independent* random variables, whereas the process ε_t is merely a sequence of *uncorrelated* variables. As far as second-order properties (such as autocovariance and spectral density functions) are concerned, e_t and ε_t have identical properties; in particular, both processes have uniform ("white") spectra. Nevertheless, they may differ substantially in other respects. In fact, we cannot strictly refer to (2.4.2) as a "model", since here the full probabilistic properties of $\{\varepsilon_t\}$ are, to a large extent, unspecified, and consequently it does not determine the full probabilistic properties of $\{X_t\}$. To achieve this, we must specify the full probabilistic structure of $\{\varepsilon_t\}$—as we would, for example, if we said that the $\{\varepsilon_t\}$ were strictly independent.

Gaussian processes

If X_t is a Gaussian process, then (assuming (2.4.2) is invertible) ε_t is also Gaussian, and being an uncorrelated process it must also be an independent process. Hence, in this special case the distinction between the processes ε_t and e_t disappears, and any stationary Gaussian process with a purely continuous spectrum satisfying (1.4.7) can be described by a linear model of the form (2.1.10).

Returning now to the general case of non-Gaussian processes, let us consider the problem of forecasting a future value, given observations up to time t. In the case of the strictly independent process, e_t, the past contains no information on the future, and hence the best forecast of any future value of e_t is simply its (unconditional) mean; namely zero. For the uncorrelated process, ε_t, it is still true that if we restrict attention to forecasts which are linear functions of past observations, then, in this sense, the past contains no information on the future. However, the past may well contain useful information on the future values if we extend the discussion to forecasts which are non-linear functions of the observations.

The following example illustrates the point very clearly. Consider the process η_t defined by,

$$\eta_t = e_t + \beta e_{t-1} e_{t-2} \tag{2.4.3}$$

where e_t is a strictly independent process with zero mean and constant variance. It follows immediately that η_t has zero mean, constant variance, and autocovariance function given by,

$$E[\eta_t \eta_{t+s}] = E[e_t e_{t+s} + \beta e_{t-1} e_{t-2} e_{t+s} + \beta e_t e_{t+s-1} e_{t+s-2} + \beta^2 e_{t-1} e_{t-2} e_{t+s-1} e_{t+s-2}].$$

For all $s \neq 0$ each of the above terms has zero expectation. (For example, in the second and third terms there is always at least one e whose time point differs from the other two—and a similar argument holds for the fourth term.) Thus, η_t is an uncorrelated process, and, as far as its second-order properties are concerned, behaves just like a white noise process. However, given observations up to time t one can clearly construct a non-trivial forecast of η_{t+1}. Specifically, if we adopt the mean square error criterion—in which case the optimal forecast of a future observation, η_{t+h}, is its conditional expectation, $\tilde{\eta}_{t+h} = E[\eta_{t+h} \mid \eta_t, \eta_{t-1}, \ldots]$ (see *Spec. Anal.*, Chapter 10), we find

$$\tilde{\eta}_{t+1} = \beta e_t e_{t-1}. \tag{2.4.4}$$

(In order to express $\tilde{\eta}_{t+1}$ in terms of past η_ts we would, of course, have to invert (2.4.3) so as to express each e_t as a function of η_t, $\eta_{t-1}, \ldots$.) As noted by Granger and Andersen (1978), if a process η_t of the above form was obtained as the residuals from a more general model, all the conventional tests for "white noise" based on the behaviour of the autocovariance or autocorrelation function would confirm that the residuals were, in fact, white noise, and hence that there was no further model structure left to fit. However, as we have seen, one could certainly exploit the non-linear structure in the η_t process in order to improve forecasts of the original series.

The above example shows that one can indeed have a stationary process which does not conform to a general linear model of the form (2.1.10) and it follows, *a fortiori*, that many types of non-stationary processes would also fall outside the domain of linear models.

Chapter 3

General Non-linear Models

3.1 VOLTERRA SERIES EXPANSIONS

In Section 2.1 we discussed the general form of a time series model, and in equation (2.1.1) we set down the most general form of model which we could envisage. In this formulation we seek a function h of past, present, and future values which reduces the given series $\{X_t\}$ to a zero mean strict white noise process $\{e_t\}$. However, just as with linear models, we will often require the general model to be "non-anticipative"—in the sense that X_t should depend only on past values and not on future ones. With this restriction (2.1.1) takes the form,

$$h(X_t, X_{t-1}, X_{t-2}, \dots) = e_t, \tag{3.1.1}$$

which represents the general non-anticipative, non-linear model. Suppose now that the function h is such that the model (3.1.1) is "invertible", i.e. that (3.1.1) may be "solved" so as to express each X_t as some (non-linear) function of $e_t, e_{t-1}, e_{t-2}, \dots$. Then we may write (3.1.1) in the equivalent form, say,

$$X_t = h'(e_t, e_{t-1}, e_{t-2}, \dots). \tag{3.1.2}$$

Without imposing further conditions on h' (or h) there is little more we can say about the general non-linear model. However, if we now assume that h' is sufficiently well-behaved so that it can be expanded in a Taylor series about some fixed point—say about the point $\mathbf{0} = (0, 0, 0, \dots)$, then we can formally expand the RHS of (3.1.2) and write,

$$X_t = \mu + \sum_{u=0}^{\infty} g_u e_{t-u} + \sum_{u=0}^{\infty} \sum_{v=0}^{\infty} g_{uv} e_{t-u} e_{t-v} + \sum_{u=0}^{\infty} \sum_{v=0}^{\infty} \sum_{w=0}^{\infty} g_{uvw} e_{t-u} e_{t-v} e_{t-w} + \cdots, \tag{3.1.3}$$

where we have written

$$\mu = h'(\mathbf{0}), \qquad g_u = \left(\frac{\partial h'}{\partial e_{t-u}}\right)_{\mathbf{0}}, \qquad g_{uv} = \left(\frac{\partial^2 h'}{\partial e_{t-u}\partial e_{t-v}}\right)_{\mathbf{0}},$$

$$g_{uvw} = \left(\frac{\partial^3 h'}{\partial \varepsilon_{t-u}\partial \varepsilon_{t-v}\partial \varepsilon_{t-w}}\right)_{\mathbf{0}}, \quad \text{etc.}$$

This expansion is known as a (discrete time) *Volterra series* (cf. Volterra, 1959), and it provides an important type of representation for non-linear models. At this stage we may regard (3.1.3) purely as a formal expansion of (3.1.2), but if we wish to investigate the statistical properties of the model (3.1.3) then clearly we must specify conditions on the sequences $\{g_u\}, \{g_{uv}\}, \{g_{uvw}\}, \ldots$ which ensure that the RHS of (3.1.3) converges to a well-defined random variable. These sequences are called the *kernels* of the Volterra series. Wiener (1958) used this type of expansion as the basis of his pioneering study of non-linear systems, although he worked in continuous time and his main objective was to construct transformations of these series in which successive terms are orthogonal—rather like the use of orthogonal polynomials in classical polynomial regression analysis. (A Volterra series expansion in continuous time is formed by the usual device of replacing the multiple summations in each term in (3.1.3) by multiple integrals.)

If we now compare (3.1.3) with (2.1.8) we see that the first term in (3.1.3) is identical with the linear model (2.1.8). Continuing our analogy with polynomial regression we may then describe the successive terms as the *quadratic, cubic, quartic, . . .* , components. (It must be noted, however, that the analogy with polynomial regression is at best a superficial one, and should not be pursued too far.)

Generalized transfer functions

Wiener's study of non-linear models was concerned mainly with the case of physical systems where both the input and output processes are observable (as opposed to the case of time series models where the "input" is an unobservable white noise process). Let U_t, X_t, denote respectively the input and output of such a system. Replacing e_t by U_t in (3.1.3) (and dropping the constant term), a non-linear relationship between $\{X_t\}$ and $\{U_t\}$ may be expressed by a Volterra series of the form,

$$X_t = \sum_{k=0}^{\infty} g_u U_{t-u} + \sum_{u=0}^{\infty}\sum_{v=0}^{\infty} g_{uv} U_{t-u} U_{t-v} + \sum_{u=0}^{\infty}\sum_{v=0}^{\infty}\sum_{w=0}^{\infty} g_{uvw} U_{t-u} U_{t-v} U_{t-w} + \cdots. \tag{3.1.4}$$

In the special case of a linear system (when only the first term in (3.1.4) is present) we know that the model is completely characterized by the system's *transfer function,*

$$\Gamma_1(\omega) = \sum_{u=0}^{\infty} g_u \, e^{-i\omega u}$$

as discussed in Section 1.4. Here, the importance of the transfer function is due to the fact that linear systems possess two fundamental properties, namely,

(1) if $U_t = A_0 \, e^{i\theta_0 t}$, $X_t = \Gamma_1(\theta_0) A \, e^{i\theta_0 t}$, i.e. if the input is a sine wave of frequency θ_0 the output is a *sine wave of the same frequency* but with the amplitude scaled by $|\Gamma_1(\theta_0)|$, and the phase shifted by $\arg\{\Gamma_1(\theta_0)\}$;

(2) if $U_t = A_0 \, e^{i\theta_0 t} + A_1 \, e^{i\theta_1 t}$, $X_t = \Gamma_1(\theta_0) A_0 \, e^{i\theta_0 t} + \Gamma_1(\theta_1) A_1 \, e^{i\theta_1 t}$, i.e. the total output is the sum of the outputs corresponding to the individual frequency components of the input. This property clearly extends to the case where the input is the sum of any number of individual frequency components, and is known as the *principle of superposition.* As a consequence of this result we may determine completely the response of a linear system to an arbitrary input simply by knowing the response of the system to sine wave inputs at all frequencies—and this is precisely the information contained in the transfer function, $\Gamma_1(\omega)$.

However, *neither property* (1) *nor property* (2) *holds for non-linear systems.* As we shall see, if we input a sine wave of frequency θ_0 into a non-linear system the output will, in general, contain not only a component at the same frequency θ_0, but also components at frequencies $2\theta_0, 3\theta_0, 4\theta_0, \ldots$. This phenomenon, called *frequency multiplication,* is well known in connection with, for example, audio amplifiers where the presence of the additional frequency components is known as *harmonic distortion.* Also, if we input a linear combination of sine waves at frequencies θ_0, θ_1, the output will contain components at frequencies θ_0, θ_1, $(\theta_0 + \theta_1)$, and all harmonics of these frequencies. (In the case of a real-valued input, say $U_t = A_0 \sin \theta_0 t + A_1 \sin \theta_1 t$, the output will contain components also at frequencies $(\theta_0 \pm \theta_1)$ and their harmonics.) This feature is again well known in audio amplifier design, where it is termed *intermodulation distortion.* It is also one of the basic theoretical results which underlie the design of the "mixer" stage in superheterodyne radio receivers.

Since neither (1) nor (2) hold for general non-linear systems we are forced to conclude that in this case *there is no such thing as a "transfer" function,* i.e. there is no single function of frequency which completely describes the response of the system to an arbitrary input. Rather, we have to introduce an *infinite sequence* of functions, namely

$$\Gamma_1(\omega_1) = \sum_{u=0}^{\infty} g_u \, e^{-i\omega_1 u},$$

$$\Gamma_2(\omega_1, \omega_2) = \sum_{u=0}^{\infty} \sum_{v=0}^{\infty} g_{uv} \, e^{-i(\omega_1 u + \omega_2 v)} \tag{3.1.5}$$

$$\Gamma_3(\omega_1, \omega_2, \omega_3) = \sum_{u=0}^{\infty} \sum_{v=0}^{\infty} \sum_{w=0}^{\infty} g_{uvw} \, e^{-i(\omega_1 u + \omega_2 v + \omega_3 w)}$$

$$\vdots$$

These are called the *generalized transfer functions* of the system. Clearly, if we know the complete sequence of functions, $\Gamma_1, \Gamma_2, \Gamma_3, \ldots$, then in principle we can invert the transforms (3.1.5), recover the sequences $\{g_u\}, \{g_{uv}\}, \{g_{uvw}\}, \ldots$, and thus completely specify the system. (Note that, without loss of generality, we may take each sequence, $g_{uvw}, \ldots$, to be a symmetric function of $u, v, w, \ldots$, since if this were not so we could rewrite each of the terms on the RHS of (3.1.4) so as to make the coefficients symmetric.)

To obtain insight into the physical interpretation of the generalized transfer functions, let us consider the case of a stationary stochastic input, i.e. take U_t to be a stationary process with spectral representation,

$$U_t = \int_{-\pi}^{\pi} e^{it\omega} \, dZ_U(\omega),$$

say. Then from (3.1.4) we may write the output as,

$$\begin{aligned} X_t = &\int_{-\pi}^{\pi} e^{it\omega_1}\Gamma_1(\omega_1) \, dZ_U(\omega_1) \\ &+ \int_{-\pi}^{\pi} \int_{\pi}^{\pi} e^{it(\omega_1+\omega_2)}\Gamma_2(\omega_1, \omega_2) \, dZ_U(\omega_1) \, dZ_U(\omega_2) \\ &+ \int_{-\pi}^{\pi} \int_{-\pi}^{\pi} \int_{-\pi}^{\pi} e^{it(\omega_1+\omega_2+\omega_3)}\Gamma_3(\omega_1, \omega_2, \omega_3) \\ &\qquad dZ_U(\omega_1) \, dZ_U(\omega_2) \, dZ_U(\omega_3) \\ &+ \cdots. \end{aligned} \tag{3.1.6}$$

We thus see that $\Gamma_1(\omega_1)$ has a similar interpretation to that of the conventional transfer function of a linear model, while

$\{\Gamma_2(\omega_1, \omega_2)\, dZ_U(\omega_1)\, dZ_U(\omega_2)\}$ represents the contribution of the components with frequencies ω_1, ω_2, in U_t to the component with frequency $(\omega_1+\omega_2)$ in X_t, etc. In particular, if U_t consists of just a single sine wave, say $U_t = A_0\, e^{i\theta_0 t}$, then $dZ_U(\omega) = A_0$, $\omega = \theta_0$, and is zero otherwise, and hence (3.1.6) becomes

$$X_t = A_0\Gamma_1(\theta_0)\, e^{i\theta_0 t} + A_0^2\Gamma_2(\theta_0, \theta_0)\, e^{2i\theta_0 t} + A_0^3\Gamma_3(\theta_0, \theta_0, \theta_0)\, e^{3i\theta_0 t} + \cdots. \quad (3.1.7)$$

The output thus consists of components with frequencies $\theta_0, 2\theta_0, 3\theta_0, \ldots$, etc., as may be verified by substituting $U_t = A_0\, e^{i\theta_0 t}$ directly in (3.1.4). This result confirms the remarks made above about the phenomenon of "frequency multiplication" in non-linear systems.

Suppose now that the input consists of two sine waves, i.e. $U_t = A_0\, e^{i\theta_0 t} + A_1\, e^{i\theta_1 t}$, so that $dZ_U(\omega) = A_0$, $\omega = \theta_0$, $dZ_U(\omega) = A_1$, $\omega = \theta_1$, and is zero otherwise. Then (3.1.6) gives

$$\begin{aligned} X_t = {} & A_0\Gamma_1(\theta_0)\, e^{i\theta_0 t} + A_1\Gamma_1(\theta_1)\, e^{i\theta_1 t} \\ & + A_0^2\Gamma_2(\theta_0, \theta_0)\, e^{2i\theta_0 t} + A_0A_1\Gamma_2(\theta_1, \theta_0)\, e^{i(\theta_0+\theta_1)t} \\ & + A_1A_0\Gamma_2(\theta_0, \theta_1)\, e^{i(\theta_0+\theta_1)t} + A_1^2\Gamma_2(\theta_1, \theta_1)\, e^{2i\theta_1 t} \\ & + \cdots \end{aligned} \quad (3.1.8)$$

thus confirming that the output of a non-linear system consists of the input frequencies and all their harmonics, together with the sum (and, in the case of real-valued inputs, the difference) of the input frequencies, and all their harmonics. (Note that a real-valued input component may be written, for example, as $A_0(e^{i\theta_0 t} + e^{-i\theta_0 t})$.)

Continuous time systems

When the input and output are both recorded as continuous functions of time, the Volterra series relationship takes the form,

$$\begin{aligned} X_t = {} & \int_0^\infty g(u)U(t-u)\,du + \int_0^\infty \int_0^\infty g(u, v)U(t-u)U(t-v)\, du\, dv \\ & + \int_0^\infty \int_0^\infty \int_0^\infty g(u, v, w)U(t-u)U(t-v)U(t-w)\, du\, dv\, dw \\ & + \cdots. \end{aligned} \quad (3.1.9)$$

The generalized transfer functions are now defined by an obvious analogy with (3.1.5), namely

$$\Gamma_1(\omega_1) = \int_0^\infty g(u)\, e^{-i\omega_1 u}\, du$$

$$\Gamma_2(\omega_1, \omega_2) = \int_0^\infty \int_0^\infty g(u, v)\, e^{-i\omega_1 u - i\omega_2 v}\, du\, dv \tag{3.1.10}$$

Identification of Volterra series

Within the context of non-linear systems there is a vast literature on the theoretical properties of Volterra series—see, for example, the comprehensive review by Brillinger (1970). By contrast, there are relatively few results on the statistical identification of Volterra series from input/output data. The basic problem may be stated thus; given a finite sample of observations on (U_t, X_t), we wish to estimate the kernels, $\{g_u\}, \{g_{uv}\}, \ldots$, or equivalently, the generalized transfer functions, $\Gamma_1(\omega_1), \Gamma_2(\omega_1, \omega_2) \ldots$. At first this may seem an impossible task since we are dealing with an infinite set of sequences, with each sequence containing an infinite number of unknown parameters, and we cannot possibly hope to estimate these from a finite sample of input/output data. However, this difficulty is no different, in principle, from that encountered in the study of linear systems, since even the "linear" term involves an infinite number of unknown parameters, namely the sequence $\{g_u\}$. In the linear case we obviate this difficulty by assuming either (i) that $\Gamma_1(\omega)$ (the Fourier transform of the $\{g_u\}$) has certain "smoothness" properties which allow the estimation of this function from finite samples, or (ii) that the $\{g_u\}$ may be expressed as known functions of some relatively small number of other parameters. Assumption (i) leads to the estimation of $\Gamma_1(\omega)$ via cross-spectral analysis (see *Spec. Anal.*, p. 786), while (ii) is usually effected by assuming that $\Gamma_1(\omega)$ is a rational function of $e^{-i\omega}$—leading to the construction of a "finite parameter" ARMA-type relationship between X_t and U_t. For non-linear systems the generalization of (i) is obtained by assuming that all the generalized transfer functions have the required smoothness properties, while the extension of (ii) is obtained by constructing special types of finite parameter non-linear models—as discussed in Chapter 4. For the present we will follow the "frequency domain" approach, and consider the estimation of the generalized transfer functions.

Pursuing the analogy between (3.1.4) and a polynomial regression model, it is evident that we cannot hope to estimate all the generalized transfer functions using only second-order properties of U_t and X_t. Clearly, if we wish to fit the first m terms in (3.1.4) we will require all the joint moments up to order $(m+1)$ of U_t and X_t. Following the method used for estimating transfer functions of linear system, this suggests that it would be useful to introduce Fourier transforms of these higher-order moments, or equivalently, of the higher-order cumulants, and we develop the relevant theory in the following section.

3.2 POLYSPECTRA

Let $\{X_t\}$ be a process stationary up to order k, and let $C(s_1, s_2, \ldots, s_{k-1})$ denote the joint cumulant of order k of the set of random variables $\{X_t, X_{t+s_1}, \ldots, X_{t+s_{k-1}}\}$, i.e. $C(s_1, s_2, \ldots, s_{k-1})$ is the coefficient of $(z_1 z_2 \ldots z_k)$ in the expansion of the cumulant generating function

$$K(z_1, z_2, \ldots, z_k) = \log_e[E[\exp\{z_1 X_t + z_2 X_{t+s_1} + \cdots + z_k X_{t+s_{k-1}}\}]].$$

Note that, by the stationarity condition, $C(s_1, s_2, \ldots, s_{k-1})$ does not depend on t.

The *kth-order polyspectrum* (or *kth-order cumulant spectrum*) is defined by

$$h_k(\omega_1, \omega_2, \ldots, \omega_{k-1}) = \left(\frac{1}{2\pi}\right)^{k-1} \sum_{s_1=-\infty}^{\infty} \cdots \sum_{s_{k-1}=-\infty}^{\infty} C(s_1, s_2, \ldots, s_{k-1}) \exp[-i(\omega_1 s_1 + \cdots + \omega_{k-1} s_{k-1})] \tag{3.2.1}$$

assuming, of course, that the above Fourier transform exists—a sufficient condition being that,

$$\sum_{s_1} \cdots \sum_{s_{k-1}} |C(s_1, s_2, \ldots, s_{k-1})| < \infty.$$

Now the second-order cumulant $C(s_1)$ is just $\text{cov}(X_t, X_{t+s_1})$, and hence it follows that the second-order polyspectrum is exactly the same as the conventional power spectrum, i.e. $h_2(\omega) \equiv h(\omega)$. The third-order cumulant $C(s_1, s_2)$ is the same as the third-order moment about the mean, i.e.

$$C(s_1, s_2) = E[\{X_t - \mu_x\}\{X_{t+s_1} - \mu_x\}\{X_{t+s_2} - \mu_x\}] = \mu(s_1, s_2), \quad \text{say}$$

(where $\mu_x = E[X_t]$), and hence the third-order polyspectrum may be written,

$$h_3(\omega_1, \omega_2) = \left(\frac{1}{2\pi}\right)^2 \sum_{s_1=-\infty}^{\infty} \sum_{s_2=-\infty}^{\infty} \mu(s_1, s_2)\, e^{-i(\omega_1 s_1 + \omega_2 s_2)},$$

$$-\pi \le \omega_1 \le \pi, \; -\pi \le \omega_2 \le \pi. \tag{3.2.2}$$

Inverting this transform we obtain,

$$\mu(s_1, s_2) = \int_{-\pi}^{\pi} \int_{-\pi}^{\pi} e^{i(\omega_1 s_1 + \omega_2 s_2)} h_3(\omega_1, \omega_2)\, d\omega_1\, d\omega_2. \tag{3.2.3}$$

The function $h_3(\omega_1, \omega_2)$ is often called the *bispectrum*, (following Tukey, 1959), and the fourth-order polyspectrum, $h_4(\omega_1, \omega_2, \omega_3)$, is called the *trispectrum*. Introducing the spectral representation for the zero-mean process $\{X_t - \mu_x\}$, i.e. writing

$$X_t - \mu_x = \int_{-\pi}^{\pi} e^{it\omega}\, dZ_x(\omega),$$

we see that $\mu(s_1, s_2)$ can be written as

$$\mu(s_1, s_2) = \int_{-\pi}^{\pi}\int_{-\pi}^{\pi}\int_{-\pi}^{\pi} e^{it(\omega_1+\omega_2+\omega_3)}\, e^{i(s_1\omega_1+s_2\omega_2)} E[dZ_x(\omega_1)\, dZ_x(\omega_2)\, dZ_x(\omega_3)]. \tag{3.2.4}$$

Since the LHS is a function of s_1, s_2, only (and does not depend on t) the RHS must similarly be a function of s_1, s_2, only. Hence the expression $E[dZ_x(\omega_1)\, dZ_x(\omega_2)\, dZ(\omega_3)]$ must vanish except along the plane $\omega_1 + \omega_2 + \omega_3 = O(\text{mod}\, 2\pi)$. Comparing (3.2.3), (3.2.4), we may now write the bispectrum in the alternative form

$$h_3(\omega_1, \omega_2)\, d\omega_1\, d\omega_2 = E[dZ_x(\omega_1)\, dZ_x(\omega_2)\, dZ_x(-\omega_1 - \omega_2)]. \tag{3.2.5}$$

The most important property of polyspectra is that *all polyspectra of higher than second order vanish when* $\{X_t\}$ *is a Gaussian process.* This follows from the well-known result that all joint cumulants of higher than second order vanish for multivariate normal distributions. Hence, the bispectrum, trispectrum, and all higher-order polyspectra are identically zero if X_t is a Gaussian process. These higher-order spectra may thus be regarded as measures of the departure of the process from Gaussianity.

Cross-polyspectra

Given two processes, X_t, Y_t, we may similarly define *cross-polyspectra* of various orders, assuming, of course, that (X_t, Y_t) are jointly stationary up to the appropriate order. For example, the *third-order cross-polyspectrum* (or *cross bispectrum*) is defined by,

$$h_{XXY}(\omega_1, \omega_2) = \left(\frac{1}{2\pi}\right)^2 \sum_{s_1=-\infty}^{\infty} \sum_{s_2=-\infty}^{\infty} \mu_{XXY}(s_1 s_2)\, e^{-i(\omega_1 s_1 + \omega_2 s_2)},$$

$$-\pi \leq \omega_1 \leq \pi, \qquad -\pi \leq \omega_2 \leq \pi. \tag{3.2.6}$$

where

$$\mu_{XXY}(s_1, s_2) = E[\{Y_t - \mu_Y\}\{X_{t+s_1} - \mu_X\}\{X_{t+s_2} - \mu_X\}],$$

and the inversion of (3.2.6) gives

$$\mu_{XXY}(s_1, s_2) = \int_{-\pi}^{\pi}\int_{-\pi}^{\pi} e^{i(\omega_1 s_1 + \omega_2 s_2)} h_{XXY}(\omega_1, \omega_2)\, d\omega_1\, d\omega_2. \quad (3.2.7)$$

Introducing the spectral representation for $\{Y_t - \mu_Y\}$, namely

$$Y_t - \mu_Y = \int_{-\pi}^{\pi} e^{it\omega}\, dZ_y(\omega),$$

a similar argument to that given above shows that $h_{XXY}(\omega_1, \omega_2)$ may be written as

$$h_{XXY}(\omega_1, \omega_2)\, d\omega_1\, d\omega_2 = E[dZ_x(\omega_1)\, dZ_x(\omega_2)\, dZ_y(-\omega_1 - \omega_2)]. \quad (3.2.8)$$

The second-order cross-polyspectrum is, of course, identical with the conventional cross-spectrum between X_t and Y_t.

The idea of constructing the Fourier transforms of higher-order cumulants was suggested by Kolmogorov, and polyspectra were introduced by Shiryaev (1960). Brillinger (1965) and Brillinger and Rosenblatt (1967a, 1967b) have given a comprehensive account of the theoretical properties of polyspectra and discuss also their estimation from sample records. Bispectra are discussed by Tukey (1959) and Akaike (1966), and an application of bispectral analysis to the study of ocean waves is described by Hasselman *et al.* (1963). Applications to tides are given in Cartwright (1968), and to turbulence by Lii *et al.* (1976) and Helland *et al.* (1979).

3.2.1 Estimation of Polyspectra

The basic ideas involved in the estimation of polyspectra are essentially the same as those used in the estimation of second-order spectra, a detailed account of which is given in *Spec. Anal.*, Chapters 6 and 7. In essence, we start with the expression for the polyspectrum as the expectation of products of the $\{dZ_x(\omega)\}$ (see, for example, (3.2.5)), and then replace the $\{dZ_x(\omega)\}$ by their corresponding sample versions, called the "finite Fourier transforms". The relevant products of these finite Fourier transforms are then "smoothed" by averaging over neighbouring sets of frequencies to produce estimates of the required polyspectrum.

Suppose that we observe the process $\{X_t\}$ over the period $t = 1, \ldots, N$, yielding N observations $X_1, X_2, \ldots, X_N$. The *finite Fourier transform* is defined by,

$$d_x(\omega) = \sum_{t=1}^{N} (X_t - \bar{X})\, e^{-i\omega t}, \qquad -\pi \le \omega \le \pi, \quad (3.2.9)$$

$\bar{X}=(1/N)\{\sum_{t=1}^{N} X_t\}$ denoting the sample mean. (For a discussion of the relationship between $d_x(\omega)$ and $dZ_x(\omega)$ see *Spec. Anal.*, p. 419.) We now form the *kth-order periodogram*, defined by,

$$I_N(\omega_1,\omega_2,\ldots,\omega_{k-1})=\frac{1}{N(2\pi)^{k-1}}\prod_{j=1}^{k} d_x(\omega_j), \qquad \sum_{j=1}^{k}\omega_j=0 \pmod{2\pi}. \tag{3.2.10}$$

Then $I_N(\omega_1,\omega_2,\ldots,\omega_{k-1})$ is an asymptotically unbiased estimate of $h_k(\omega_1,\omega_2,\ldots,\omega_{k-1})$, but is not a consistent estimate (Brillinger and Rosenblatt, 1967a, 1967b). To construct a consistent estimate we now "smooth" the function $I_N(\omega_1,\omega_2,\ldots,\omega_{k-1})$ by introducing a "weight function" which becomes increasingly more concentrated as the sample size $N\to\infty$. Thus, let $K(\theta_1,\ldots,\theta_k)$ be a function defined on the plane $\sum_{j=1}^{k}\theta_j=0 \pmod{2\pi}$, and satisfying,

(i) $K(\theta_1,\ldots,\theta_k)=K(-\theta_1,\ldots,-\theta_k)$,

(ii) $\displaystyle\int_{-\infty}^{\infty}\cdots\int_{-\infty}^{\infty} K(\theta_1,\ldots,\theta_k)\delta\left(\sum_{j=1}^{k}\theta_j\right) d\theta_1,\ldots,d\theta_k=1$, and

(iii) $\displaystyle\int_{-\infty}^{\infty}\cdots\int_{-\infty}^{\infty} K^2(\theta_1,\ldots,\theta_k)\delta\left(\sum_{j=1}^{k}\theta_j\right) d\theta_1,\ldots,d\theta_k<\infty$,

$\delta(\cdot)$ being the Dirac δ-function. Now set

$$W_N(\theta_1,\ldots,\theta_k)=M^{k-1}K(M\theta_1,\ldots,M\theta_k),$$

where the integer $M(=M(N))$ is chosen so that $M\to\infty$ as $N\to\infty$, but $M^{k-1}/N\to 0$ as $N\to\infty$. The function W_N is called the *spectral window* and K is called the *spectral window generator*. The "smoothed" estimate is then given by

$$\hat{h}_k(\omega_1,\ldots,\omega_{k-1})=\int_{-\pi}^{\pi}\cdots\int_{-\pi}^{\pi} W_N(\theta_1-\omega_1,\ldots,\theta_k-\omega_k)$$
$$\times\,\delta(\textstyle\sum\theta_j)I_N(\theta_1,\ldots,\theta_k)\,d\theta_1,\ldots,d\theta_k, \tag{3.2.11}$$

with $\omega_k=-(\omega_1+\omega_2+\cdots+\omega_{k-1})$. (Alternatively, the RHS of (3.2.11) may be replaced by a k-fold summation over the discrete grid of frequencies $(2\pi p_1/N,\ldots,2\pi p_k/N)$—as in the second-order case.) A comprehensive discussion of the asymptotic sampling properties of these estimates is given by Brillinger and Rosenblatt (1967a, 1967b).

To illustrate the above ideas, we now consider in a more detailed form the estimation of the *bispectrum*, $h_3(\omega_1,\omega_2)$. We first note that from (3.2.2), $h_3(\omega_1,\omega_2)$ may be expressed as the Fourier transform of the third-order moments, $\mu(s_1, s_2)$. Now for real-valued processes it follows from the definition that the function $\mu(s_1, s_2)$ satisfies the symmetry relations,

$$\mu(s_1, s_2) = \mu(s_2, s_1) = \mu(-s_1, s_2 - s_1) = \mu(s_1 - s_2, -s_2), \quad (3.2.12)$$

and hence $h_3(\omega_1, \omega_2)$ satisfies the corresponding symmetry relations,

$$h_3(\omega_1, \omega_2) = h_3(\omega_2, \omega_1) = h_3(\omega_1, -\omega_1 - \omega_2) = h_3^*(-\omega_1, -\omega_2). \quad (3.2.13)$$

The form of $h_3(\omega_1, \omega_2)$ over the complete (ω_1, ω_2) plane is thus completely determined by its values in any one of 12 sectors of this plane—as described by Subba Rao and Gabr (1984, p. 13). Now consider the estimate of $h_3(\omega_1, \omega_2)$ derived from the general form (3.2.11). Here, we may express the third-order periodogram more explicitly as a function of just ω_1, ω_2, by setting $k = 3$ and $\omega_3 = -(\omega_1 + \omega_2)$ in (3.2.10), and we may also write the weight functions K, W_N, purely as functions of θ_1, θ_2, in which case W_N then satisfies the symmetry relations,

$$W_N(\theta_1, \theta_2) = W_N(\theta_2, \theta_1) = W_N(\theta_1, -\theta_1 - \theta_2) = W_N(-\theta_1 - \theta_2, \theta_2), \quad (3.2.14)$$

and similarly for $K(\theta_1, \theta_2)$.

The estimate of $h_3(\omega_1, \omega_2)$ may now be written

$$\hat{h}_3(\omega_1, \omega_2) = \int_{-\pi}^{\pi} \int_{-\pi}^{\pi} W_N(\theta_1 - \omega_1, \theta_2 - \omega_2) I_N(\theta_1, \theta_2)\, d\theta_1\, d\theta_2. \quad (3.2.15)$$

Alternatively, $\hat{h}_3(\omega_1, \omega_2)$ may be written in the algebraically equivalent form,

$$\hat{h}_3(\omega_1, \omega_2) = \frac{1}{(2\pi)^2} \sum_{s_1=-(N-1)}^{(N-1)} \sum_{s_2=-(N-1)}^{(N-1)} k\left(\frac{s_1}{M}, \frac{s_2}{M}\right) \hat{\mu}(s_1, s_2)\, e^{-is_1\omega_1 - is_2\omega_2} \quad (3.2.16)$$

where

$$\hat{\mu}(s_1, s_2) = \frac{1}{N} \sum_{t=1}^{N-\tau} (X_t - \bar{X})(X_{t+s_1} - \bar{X})(X_{t+s_2} - \bar{X}),$$

$$s_1 \geq 0,\ s_2 \geq 0, \qquad \tau = \max(s_1, s_2), \quad (3.2.17)$$

is the obvious sample version of $\mu(s_1, s_2)$, and $k(s_1, s_2)$ is the inverse Fourier transform of $K(\theta_1, \theta_2)$, namely

$$k(s_1, s_2) = \int_{-\infty}^{\infty} \int_{-\infty}^{\infty} e^{is_1\theta_1 + is_2\theta_2} K(\theta_1, \theta_2)\, d\theta_1\, d\theta_2. \quad (3.2.18)$$

(The function $k(s_1, s_2)$ is sometimes called the *covariance lag window generator.*) The expression (3.2.16) is an obvious sample analogue of (3.2.2), and the algebraic equivalence between (3.2.15) and (3.2.16) is established by similar arguments to those used for the corresponding result in the second-order case—for which, see *Spec. Anal.*, p. 435.

Subba Rao and Gabr (1984) give a thorough and detailed discussion of the estimation of bispectra, and provide a variety of numerical illustrations of this procedure. Subba Rao and Gabr show, in particular, that under general conditions on $\mu(s_1, s_2)$ and $K(\theta_1, \theta_2)$, the bias $b(\omega_1, \omega_2) = [E\{\hat{h}_3(\omega_1, \omega_2)\} - h_3(\omega_1, \omega_2)]$ is given by,

$$b(\omega_1, \omega_2) = \frac{B_K}{M^2} D^{(2)}(\omega_1, \omega_2) + O(1/M^3), \tag{3.2.19}$$

where

$$D^{(2)}(\omega_1, \omega_2) = \left(\frac{\partial^2}{\partial\omega_1^2} - \frac{\partial^2}{\partial\omega_1 \partial\omega_2} + \frac{\partial^2}{\partial\omega_2^2}\right) h_3(\omega_1, \omega_2),$$

and

$$B_K = -\int_{-\infty}^{\infty}\int_{-\infty}^{\infty} \theta_1\theta_2 K(\theta_1, \theta_2)\, d\theta_1\, d\theta_2.$$

Also, Brillinger and Rosenblatt (1967a) have shown that the asymptotic variance is given by,

$$\text{var}\{\hat{h}_3(\omega_1, \omega_2)\} \sim \frac{M^2}{N} \frac{V_2}{2\pi} h_3(\omega_1) h_3(\omega_2) h_3(\omega_1 + \omega_2), \qquad 0 < \omega_2 < \omega_1 < \pi, \tag{3.2.20}$$

where

$$V_2 = \int_{-\infty}^{\infty}\int_{-\infty}^{\infty} k^2(s_1, s_2)\, ds_1\, ds_2 = (2\pi)^2 \int_{-\infty}^{\infty}\int_{-\infty}^{\infty} K^2(\theta_1, \theta_2)\, d\theta_1\, d\theta_2.$$

Thus, provided $M \to \infty$ as $N \to \infty$, but $M^2/N \to 0$ as $N \to \infty$, $\hat{h}_3(\omega_1, \omega_2)$ is a consistent estimate of $h_3(\omega_1, \omega_2)$.

Note, however, that for finite N the estimate $\hat{h}_3(\omega_1, \omega_2)$ will, in general, be *biased*, the magnitude of the bias depending on the "smoothness" of the theoretical bispectrum $h_3(\omega_1, \omega_2)$ over the effective "bandwidth" of the window W_N. In fact, the whole technique of "smoothing windows" depends for its effectiveness on the assumption that the function to be estimated has some degree of smoothness.

Choice of window

The choice of the specific form of $k(s_1, s_2)$ or $K(\theta_1, \theta_2)$ is to a large extent arbitrary (subject to the conditions (i), (ii), (iii), noted above), a feature which is once again similar to that encountered in the second-order case. In the latter context there are a number of special windows which have gained popularity—such as the Bartlett, Bartlett-Priestley, Daniell, Parzen, and Tukey-Hamming forms. Here both k and K are one-dimensional functions, and, for example, for the Bartlett window,

$$k(s)=\begin{cases}1-|s|, & |s|\le 1\\ 0, & |s|>1,\end{cases} \tag{3.2.21}$$

$$K(\theta)=\frac{1}{2\pi}\left(\frac{\sin(\theta/2)}{\theta/2}\right)^2,$$

while for the Parzen window,

$$k(s)=\begin{cases}1-6u^2+6|u|^3, & |u|\le\frac{1}{2},\\ 2(1-|u|^3), & \frac{1}{2}\le|u|\le 1,\\ 0, & |u|>1,\end{cases} \tag{3.2.22}$$

$$K(\theta)=\frac{3}{8\pi}\left(\frac{\sin(\theta/4)}{\theta/4}\right)^4.$$

(A detailed description of other standard windows is given in *Spec. Anal.*, pp. 437–449.)

Given a particular one-dimensional lag window, $k(s)$, it is, of course, possible to construct a two-dimensional lag window by setting

$$k(s_1, s_2)=k(s_1)k(s_2).$$

However, as pointed out by Subba Rao and Gabr (1984), this choice of $k(s_1, s_2)$ does not satisfy the required symmetry conditions. They therefore suggest that a more appropriate form is that given by setting

$$k(s_1, s_2)=k(s_1)k(s_2)k(s_1-s_2). \tag{3.2.23}$$

In particular, if $k(u)$ has "characteristic exponent" 2 (see *Spec. Anal.*, p. 459), then it is not difficult to show that with $k(s_1, s_2)$ given by (3.2.23),

$$B_K=-k''(0).$$

There are various criteria according to which one can compare the "efficiencies" of difference windows, but from the point of view of minimizing the relative mean-square error of spectral estimates it may be shown (*Spec.*

Anal., p. 569) that the optimum choice is the Bartlett–Priestley window, namely

$$k(s)=\frac{3}{(\pi s)^2}\left(\frac{\sin \pi s}{\pi s}-\cos \pi s\right),$$
$$K(\theta)=\begin{cases}\frac{3}{4\pi}(1-(\theta/\pi)^2) & |\theta|\le\pi\\ 0, & |\theta|>\pi.\end{cases} \tag{3.2.24}$$

Subba Rao and Gabr (1984) have extended this result to the bispectral case, and show that, according to the same criterion, the optimal two-dimensional window is given by

$$K^*(\theta_1,\theta_2)=\begin{cases}\frac{\sqrt{3}}{\pi^3}\left\{1-\frac{1}{\pi^2}(\theta_1^2+\theta_2^2+\theta_1\theta_2)\right\}, & \text{if } (\theta_1,\theta_2)\in G_1\\ 0, & \text{otherwise,}\end{cases} \tag{3.2.25}$$

where the region G_1 is defined by

$$\{(\theta_1,\theta_2);\ \theta_1^2+\theta_1^2+\theta_1\theta_2\le\pi^2\}.$$

The inversion of $K^*(\omega_1,\omega_2)$ to produce $k(s_1,s_2)$ is complicated, but Subba Rao and Gabr have derived an approximate expression for the corresponding lag window by replacing the ellipse G_1 by a simpler region consisting of a combination of triangles and rectangles.

3.3 ESTIMATION OF GENERALIZED TRANSFER FUNCTIONS

Consider a non-linear system whose input U_t and output X_t satisfy the Volterra series relationship (3.1.4). Suppose we wish to fit a finite number of such terms to sample input/output data. If we think back, once again, to the analogy with polynomial regression it is evident that we cannot estimate a particular kernel, $g_{uvw\ldots}$ (or equivalently, the corresponding transfer function) independently of the other terms included in the model since, in general, the various terms on the RHS of (3.1.4) will not be orthogonal. This means that we cannot hope to obtain simple analytical expressions for the generalized transfer functions in terms of the polyspectra and cross-polyspectra. There is, however, one special case when we can derive a very elegant generalization of the basic result for the transfer function of a linear system (*Spec. Anal.*, p. 672). This is the case where the RHS of (3.1.4) contains just one term, say the term of order m, and the input U_t is a Gaussian process. In this case we can determine the mth-order generalized transfer function in terms of the $(m+1)$th-order cross-polyspectrum between U_t and X_t.

Suppose, for example, that (3.1.4) contains just the quadratic term, together with the possible addition of a stationary "noise" term. Then we may write,

$$X_t = \sum_{u=0}^{\infty} \sum_{v=0}^{\infty} g_{uv} U_{t-u} U_{t-v} + N_t, \tag{3.3.1}$$

and we assume $E[U_t] = E[N_t] = 0$, and that N_t is independent of U_t. Using (3.1.6) we may write,

$$X_t = \int_{-\pi}^{\pi} \int_{-\pi}^{\pi} e^{it(\omega_1+\omega_2)} \Gamma_2(\omega_1, \omega_2)\, dZ_U(\omega_1)\, dZ_U(\omega_2) + N_t.$$

so that

$$E[X_t U_{t+s_1} U_{t+s_2}] = \int_{-\pi}^{\pi} \int_{-\pi}^{\pi} \int_{-\pi}^{\pi} \int_{-\pi}^{\pi} \exp\{it(\omega_1+\omega_2) + i(t+s_1)\omega_3 + i(t+s_2)\omega_4\}$$
$$\times \Gamma_2(\omega_1, \omega_2) E[dZ_U(\omega_1)\, dZ_U(\omega_2)\, dZ_U(\omega_3)\, dZ_U(\omega_4)].$$

Since $dZ_U(\omega)$ is complex Gaussian, the expectation of the above quadruple product can be written as

$$E[dZ_U(\omega_1)\, dZ_U(\omega_2)]E[dZ_U(\omega_3)\, dZ_U(\omega_4)]$$
$$+ E[dZ_U(\omega_1)\, dZ_U(\omega_3)]E[dZ_U(\omega_2)\, dZ_U(\omega_4)]$$
$$+ E[dZ_U(\omega_1)\, dZ_U(\omega_4)]E[dZ_U(\omega_2)\, dZ_U(\omega_3)].$$

If we now substitute this expression into the quadruple integral, and note that, for example,

$$E[dZ_U(\omega_1)\, dZ_U(\omega_2)] = 0, \qquad \omega_1 \neq -\omega_2$$
$$E[dZ_U(\omega_1)\, dZ_U(-\omega_1)] = h_U(\omega_1)\, d\omega_1,$$

$h_U(\omega_1)$ being the (second-order) spectral density function of U_t, we find,

$$E[X_t U_{t+s_1} U_{t+s_2}]$$
$$= \int_{-\pi}^{\pi} \Gamma_2(\omega_1, -\omega_1) h_U(\omega_1)\, d\omega_1 \int_{-\pi}^{\pi} e^{i(s_1-s_2)\omega_3} h_U(\omega_3)\, d\omega_3$$
$$+ \int_{-\pi}^{\pi} \int_{-\pi}^{\pi} \Gamma_2(-\omega_3, -\omega_4) e^{i(s_1\omega_3+s_2\omega_4)} h_U(\omega_3) h_U(\omega_4)\, d\omega_3\, d\omega_4$$
$$+ \int_{\pi}^{\pi} \int_{\pi}^{\pi} \Gamma_2(-\omega_4, -\omega_3) e^{i(s_1\omega_3+s_2\omega_4)} h_U(\omega_3) h_U(\omega_4)\, d\omega_3\, d\omega_4. \tag{3.3.2}$$

Now write

$$\mu_{UUX}(s_1, s_2) = E[X_t U_{t+s_1} U_{t+s_2}] - E[U_{t+s_1} U_{t+s_2}]E[X_t], \tag{3.3.3}$$

and noting that the first term in (3.3.2) is just the second term in (3.3.3), and that, without loss of generality, we may take $\Gamma_2(\omega_1, \omega_2)$ to be symmetric in ω_1, ω_2, we finally obtain,

$$\mu_{UUX}(s_1, s_2) = 2\int_{-\pi}^{\pi}\int_{-\pi}^{\pi} \Gamma_2(-\omega_3, -\omega_4)\, e^{i(s_1\omega_3+s_2\omega_4)} h_U(\omega_3) h_U(\omega_4)\, d\omega_3\, d\omega_4.$$

Comparing the above expression with (3.2.7) it now follows that the third-order cross-polyspectrum between U_t and X_t is given by (writing ω_1 for ω_3 and ω_2 for ω_4),

$$h_{UUX}(\omega_1, \omega_2) = 2\Gamma_2(-\omega_1, -\omega_2) h_U(\omega_1) h_U(\omega_2),$$

or,

$$\Gamma_2(\omega_1, \omega_2) = \frac{h_{UUX}(-\omega_1, -\omega_2)}{2h_U(\omega_1)h_U(\omega_2)}. \tag{3.3.4}$$

As a generalization of this result, Shiryaev (1960) has shown that when the Volterra series (3.1.4) contains just the mth-order term (with possible additive noise), the mth-order generalized transfer function is given by,

$$\Gamma_m(\omega_1, \omega_2, \ldots, \omega_m) = \frac{h_{UU\ldots UX}(-\omega_1, -\omega_2, \ldots, -\omega_m)}{m!\, h_U(\omega_1)h_U(\omega_2)\ldots h_U(\omega_m)}. \tag{3.3.5}$$

(See Brillinger, 1970.) Here, the numerator is the mth-order cross-polyspectrum between X_t and U_t. This result is an elegant generalization of the classical result for the transfer function of a linear system (see *Spec. Anal.*, p. 672, equation (9.2.19)) to which it reduces when $m = 1$.

If (3.1.4) does, in fact, contain just one term, (3.3.5) enables us to estimate $\Gamma_m(\omega_1, \omega_2, \ldots, \omega_m)$ from a finite number of observations on (U_t, X_t) by replacing the terms in the numerator and denominator by their respective sample estimates. At this stage we may note that conditions mentioned previously on the "smoothness" properties of the generalized transfer functions—or equivalently, of the cross-polyspectra and spectra. These "smoothness" properties are now seen to be an essential requirement in order that we may construct reasonably accurate non-parametric estimates of the numerator and denominator in (3.2.5) from samples of finite size based on "window" smoothing techniques described in Section 3.2.

If the Volterra series (3.1.4) contains a mixture of terms of different orders, then the result (3.3.5) is no longer valid. However, we can apply a similar approach if we first adopt Wiener's technique of rewriting the RHS of (3.1.4) as a sum of *orthogonal* terms. This type of orthogonal expansion requires the use of Hermite polynomials (Wiener, 1958), and the "re-parametrized" generalized transfer functions can then each be obtained via expressions of the form (3.3.5) (see Brillinger, 1970).

Quadratic systems

An example of the above approach is described by Subba Rao and Nunes (1985) in their study of the identification of quadratic systems (see also Tick, 1961). These authors consider a quadratic system of the form

$$X_t = \sum_{u=0}^{\infty} k_1(u)U_{t-u} + \sum_{u=0}^{\infty}\sum_{v=0}^{\infty} k_2(u, v)\{U_{t-u}U_{t-v} - R_U(u-v)\} + N_t \quad (3.3.6)$$

in which $\{U_t\}$ is a zero-mean Gaussian process, and $R_U(s) = E[U_tU_{t+s}]$ denotes the autocovariance function of $\{U_t\}$. In the form (3.3.6) the linear and quadratic terms are orthogonal, and as above we may assume that k_2 is a symmetric function of u, v. Given observations on (U_t, X_t), least-squares estimates $\{k_1(u)\}$, $\{k_2(u, v)\}$ are obtained by minimizing $\sum_t N_t^2$, which is equivalent, asymptotically, to minimizing

$$Q = E\left[X_t - \sum_u k_1(u)U_{t-u} - \sum_u\sum_v k_2(u, v)\{U_{t-u}U_{t-v} - R_U(u-v)\}\right]^2.$$

Differentiating Q with respect to $k_1(u)$, $k_2(u, v)$, and setting the derivatives to zero yields,

$$R_{UX}(s) = \sum_{u=0}^{\infty} k_1(u)R_U(s+u) \quad (3.3.7)$$

$$\begin{aligned} R_{UUX}(m, l) &= \sum_{u=0}^{\infty}\sum_{v=0}^{\infty} k_2(u, v) \\ &\quad \times [E(U_{t-u}U_{t-v}U_{t-m}U_{t-l}) - R_U(u-v)R_U(m-l)] \\ &= 2\sum_{u=0}^{\infty}\sum_{v=0}^{\infty} k_2(u, v)R_U(m+u)R_U(l+v) \end{aligned} \quad (3.3.8)$$

where $R_{UX}(s) = E[X_tU_{t+s}]$, $R_{UUX}(m, l) = E[X_tU_{t+m}U_{t+l}]$. Let $h_U(\omega)$ denote the spectral density function of U_t, $h_{UX}(\omega)$ the cross-spectral density function between U_t, X_t, and let

$$h_{UUX}(\omega) = \frac{1}{(2\pi)^2}\sum_{s_1=0}^{\infty}\sum_{s_2=0}^{\infty} R_{UUX}(s_1, s_2)\, e^{-i(s_1\omega_1 + s_2\omega_2)}$$

denote the cross-bispectrum between U_t and X_t. Taking Fourier transforms of both sides of (3.3.7), (3.3.8) we obtain,

$$A(\omega) = \sum_{u=0}^{\infty} k_1(u)\, e^{-i\omega u} = h_{UX}(\omega)/h_U(\omega), \quad (3.3.9)$$

and

$$B(\omega_1, \omega_2) = \sum_{u=0}^{\infty}\sum_{v=0}^{\infty} k_2(u, v)\, e^{-i(u\omega_1 + v\omega_2)} = \frac{h_{UUX}(-\omega_1, -\omega_2)}{2h_U(\omega_1)h_U(\omega_2)}, \quad (3.3.10)$$

(cf. (3.3.4)). Using the orthogonality of the linear and quadratic terms in (3.3.6), together with the above expressions for $A(\omega)$, $B(\omega_1, \omega_2)$, we may derive an expression for the spectral density function of X_t, first given by Tick (1961), namely

$$h_X(\omega) = \frac{|h_{UX}(\omega)|^2}{h_U(\omega)} + \frac{1}{2}\int_{-\pi}^{\pi} \frac{|h_{UUX}(\omega-\lambda, \lambda)|^2}{h_U(\omega-\lambda)h_U(\lambda)}\, d\lambda + h_N(\omega),$$

$h_N(\omega)$ denoting the spectral density function of the noise process, $\{N_t\}$. The above expression may be rearranged thus:

$$h_N(\omega) = h_X(\omega)\{1 - Q(\omega)\},$$

where

$$Q(\omega) = \frac{|h_{UX}(\omega)|^2}{h_U(\omega)h_X(\omega)} + \frac{1}{2h_X(\omega)}\int_{-\pi}^{\pi} \frac{|h_{UUX}(\omega-\lambda, \lambda)|^2}{h_U(\omega-\lambda)h_U(\lambda)}\, d\lambda \qquad (3.3.11)$$

is known as the *quadratic coherency coefficient.* If $Q(\omega) \equiv 1$, all ω, then clearly $N_t \equiv 0$, and $Q(\omega)$ may thus be interpreted as a measure of the goodness of fit of the quadratic model. (Note that the first term in (3.3.11) is just the (linear) squared coherency between U_t and X_t—see *Spec. Anal.*, p. 674.)

Given finite records on the input and output processes, the above results may be used to construct estimates of the linear and quadratic transfer functions by replacing the spectra, cross-spectra, and cross-bispectra in (3.3.9), (3.3.10), by their sample estimates. Subba Rao and Nunes (1985) apply this technique to various sets of simulated data, and apply it also to the gas furnace data of Box and Jenkins (1970).

3.4 TESTS FOR GAUSSIANITY AND LINEARITY

We have already noted that if a process is Gaussian all its polyspectra of higher order than the second are identically zero. Hence, if a process has a non-zero bispectrum (say), this could be due to the fact that either:

(1) the process conforms to a linear model of the form (2.1.8) but the $\{e_t\}$ are non-normal; or
(2) the process conforms to a non-linear model of the form (3.1.3) with the $\{e_t\}$ either normal or non-normal.

Following the approach of Brillinger (1965), Subba Rao and Gabr (1980, 1984) constructed two tests designed to detect:

(1) whether the process is *Gaussian*—in which case, given that it is

stationary with an absolutely continuous spectrum it must necessarily conform to a linear model; and

(2) if the process is non-Gaussian, whether it conforms to a *linear model.*

Case (1) is examined by testing the null-hypothesis that the bispectrum is zero to all frequencies, and the test statistic is constructed from the values of the estimated bispectrum over a suitable grid of points. Under the null-hypothesis the test statistic has a form similar to Hotelling's T^2 statistic. Case (2) is tested by using the result that for the linear model (2.1.8),

$$\mu(s_1, s_2) = \beta \sum_{u=-\infty}^{\infty} g_u g_{u+s_1} g_{u+s_2}, \tag{3.4.1}$$

where $\beta = E[e_t^3]$ is the third moment of the $\{e_t\}$, and $g_u = 0$, $u < 0$; this result follows immediately from the independence of the $\{e_t\}$. Taking Fourier transforms of both sides of (3.4.1) now gives

$$\begin{aligned} h_3(\omega_1, \omega_2) &= \frac{\beta}{4\pi^2} \left\{ \sum_u g_u \, e^{i(\omega_1+\omega_2)u} \right\} \left\{ \sum_{s_1} g_{u+s_1} \, e^{-i\omega_1(u+s_1)} \right\} \left\{ \sum_{s_2} g_{u+s_2} \, e^{-i\omega_2(u+s_2)} \right\} \\ &= \frac{\beta}{4\pi^2} \Gamma\{-(\omega_1+\omega_2)\}\Gamma(\omega_1)\Gamma(\omega_2). \end{aligned} \tag{3.4.2}$$

But the spectral density function of the process is given by

$$h(\omega) = \frac{\sigma_e^2}{2\pi} |\Gamma(\omega)|^2. \tag{3.4.3}$$

Hence the function

$$X(\omega_1, \omega_2) = \frac{|h_3(\omega_1, \omega_2)|^2}{h(\omega_1)h(\omega_2)h(\omega_1+\omega_2)} = \text{constant (all } \omega_1, \omega_2). \tag{3.4.4}$$

The test for linearity is based on replacing $h_3(\omega_1, \omega_2)$ and $h(\omega)$ by their sample estimates, and then testing the constancy of the sample values of $X(\omega_1, \omega_2)$ over a grid of points. The test statistic again has a form similar to Hotelling's T^2.

The detailed format of these tests is as follows.

1. Test of Gaussianity

As explained above, the test for Gaussianity is based on testing the null-hypothesis that the bispectrum is zero at all frequencies, and the test statistic is constructed from bispectral estimates over a grid of frequencies, $\{\omega_i, \omega_j\}$. The test is performed in two stages: first we examine the hypothesis $h_3(\omega_i, \omega_j) \equiv 0$ when $\{\omega_i, \omega_j\}$ lie *within* the region,

$$0 < \omega_i < 2\pi/3, \qquad \omega_i < \omega_j < \pi - \tfrac{1}{2}\omega_i. \tag{3.4.5}$$

In this region each bispectral estimate is approximately distributed as a complex normal variable, and the Subba Rao–Gabr test uses a complex analogue of Hotelling's T^2 statistic (Giri, 1965; Khatri, 1965). If the null-hypothesis is accepted at this stage, the second stage of the test then examines whether $h_3(\omega_i, \omega_j) = 0$ when $\{\omega_i, \omega_j\}$ lies on the boundaries of the region (3.4.5), or at the origin. In the latter case, the test is based on Hotelling's T^2 statistic for real variables (Anderson, 1958).

The first stage of the test proceeds as follows. We select a "course" grid of frequencies $\{\omega_i, \omega_j\}$, where

$$\omega_i = \frac{\pi i}{K}, \qquad \omega_j = \frac{\pi j}{K}, \qquad i = 1, 2, \ldots, L, j = i+1, i+2, \ldots, \gamma(i),$$

with $L = 2K/3$, $\gamma(i) = K - [i/2]$. Here, K must be chosen so that $K \ll N$, where N is the number of observations in the sample from which the bispectral estimates are computed. Note that with L and $\gamma(i)$ determined as above, each of the points $\{\omega_i, \omega_j\}$ lies within the region (3.4.5).

Let $\eta_{ij} = h_3(\omega_i, \omega_j)$, and for each i $(i = 1, 2, \ldots, L)$ define the vector,

$$\boldsymbol{\eta}_i' = (\eta_{i,i+1}, \eta_{i,i+2}, \ldots, \eta_{i,\gamma(i)}).$$

Now assemble all the η_i's into a single column vector by writing

$$\boldsymbol{\eta}' = (\boldsymbol{\eta}_1', \boldsymbol{\eta}_2', \ldots, \boldsymbol{\eta}_L'),$$

and relabel the elements of $\boldsymbol{\eta}$ by writing

$$\boldsymbol{\eta}' = (\zeta_1, \zeta_2, \ldots, \zeta_P),$$

where $P = \sum_{i=1}^{L} \{\gamma(i) - i\}$, so that for each $l (1 \le l \le P)$, $\zeta_l = \eta_{ij}$ for some i, j, satisfying $1 \le i \le L$; $(i+1) \le j \le \gamma(i)$.

We now form a set of (approximately) uncorrelated estimates of each ζ_i by constructing a "fine" grid of frequencies around each ω_i, ω_j point. Thus, for each ω_i, ω_j, set

$$\omega_{i_p} = \omega_i + \frac{pd\pi}{N}, \qquad p = -r, -(r-1), \ldots, 0, 1, \ldots, r,$$

$$\omega_{j_q} = \omega_j + \frac{qd\pi}{N}, \qquad q = -r, -(r-1), \ldots, 1, \ldots, r, \qquad (q \ne 0),$$

where the interval d is chosen so that the bispectral estimates at neighbouring points on the "fine" grid are approximately uncorrelated. (This implies that $\pi d/N$ must be greater than the bandwidth of the spectral window used in the estimation of the bispectral.) Denoting the estimated bispectrum at

$(\omega_{i_p}, \omega_{j_q})$ by $\hat{h}_3(\omega_{i_p}, \omega_{j_q})$, and assuming that the function h_3 is sufficiently smooth so as to be effectively constant over each "fine" grid, we may write

$$E[\hat{h}_3(\omega_{i_p}, \omega_{j_q})] \sim h_3(\omega_i, \omega_j), \quad \text{all } p, q.$$

Thus, the set of estimates $\hat{h}_3(\omega_{i_p}, \omega_{j_q})$ may be regarded as $n = 4r+1$ approximately uncorrelated and unbiased estimates of $h_3(\omega_i, \omega_j)$. We now form the set of bispectral estimates,

$$\left\{\hat{h}_3\left(\omega_i - \frac{r\,d\pi}{N}, \omega_j\right), \hat{h}_3\left(\omega_i - \frac{(r-1)\,d\pi}{N}, \omega_j\right), \ldots, \hat{h}_3(\omega_i, \omega_j), \ldots,\right.$$

$$\hat{h}_3\left(\omega_i + \frac{r\,d\pi}{N}, \omega_j\right), \hat{h}_3\left(\omega_i, \omega_j - \frac{r\,d\pi}{N}\right),$$

$$\left.\hat{h}_3\left(\omega_i, \omega_j - \frac{(r-1)\,d\pi}{N}\right), \ldots, \hat{h}_3\left(\omega_i, \omega_j + \frac{r\,d\pi}{N}\right)\right\}$$

in a $n \times 1$ column vector which, after relabelling as above, can be denoted by $\boldsymbol{\xi} - (\xi_1, \xi_2, \ldots, \xi_n)'$, where for each k, $\xi_k = \hat{h}_3(\omega_{i_p}, \omega_{j_q})$ for some p, q. By this device we obtain a $n \times 1$ vector of estimates for each element ζ_l of the vector $\boldsymbol{\eta}$, and the complete set of bispectral estimates may then be formed into a "data matrix", $\boldsymbol{D}$, namely

$$\boldsymbol{D} = \begin{bmatrix} \xi_{11} & \cdots & \xi_{1n} \\ \xi_{21} & \cdots & \xi_{2n} \\ \cdots & \cdots & \cdots \\ \xi_{P1} & & \xi_{Pn} \end{bmatrix} = (\boldsymbol{\xi}_{(1)}, \boldsymbol{\xi}_{(2)}, \ldots, \boldsymbol{\xi}_{(n)}), \quad \text{say.}$$

For large N, each $\boldsymbol{\xi}_{(i)}$ has a complex normal distribution with mean $\boldsymbol{\eta}$, and variance-covariance matrix $\boldsymbol{\Sigma}_\xi$, say. The maximum-likelihood estimates of $\boldsymbol{\eta}$ and $\boldsymbol{\Sigma}_\xi$ are given by,

$$\hat{\boldsymbol{\eta}} = \frac{1}{n} \sum_{i=1}^{n} \boldsymbol{\xi}_{(i)}, \qquad \hat{\boldsymbol{\Sigma}}_\xi = \frac{1}{n} \sum_{i=1}^{n} (\boldsymbol{\xi}_{(i)} - \hat{\boldsymbol{\eta}})(\hat{\boldsymbol{\xi}}_{(i)} - \hat{\boldsymbol{\eta}})^*.$$

Now under the null-hypothesis that the process is Gaussian, all bispectral ordinates are zero, and hence $\boldsymbol{\eta} = 0$. The likelihood ratio test for testing $H_0 : \boldsymbol{\eta} = 0$ against the alternative $H_1 : \boldsymbol{\eta}^* \boldsymbol{\Sigma}_\xi^{-1} \boldsymbol{\eta} > 0$, leads to the rejection of H_0 if

$$T^2 = n\hat{\boldsymbol{\eta}}^* \boldsymbol{A}^{-1} \boldsymbol{\eta} > \lambda,$$

where λ is a constant determined by the significance level of the test, and $\boldsymbol{A} = n\hat{\boldsymbol{\Sigma}}_\xi$ (see, for example, Giri, 1965; Khatri, 1965). To implement the test we may note that under H_0, the statistic

$$\mathscr{F}_1 = [2(n-P)/2P]T^2$$

is distributed as $F_{2P,2(n-P)}$.

2. Test for linearity

The test for linearity proceeds in much the same way as the previous one, except that here the test is based on the sample version of the $X(\omega_1, \omega_2)$ statistic defined by (3.4.4), evaluated over a suitable grid of frequencies. Thus, with ω_i, ω_j defined as in (1), we write

$$X_{ij} = X(\omega_i, \omega_j) = \frac{|h_3(\omega_i, \omega_j)|^2}{h(\omega_i)h(\omega_j)h(\omega_i+\omega_j)},$$

and form the $\{X_{ij}\}$ values $(0 < \omega_i < \pi, \omega_i < \omega_j < \pi)$ into a $P \times 1$ column vector, which we may write as

$$\boldsymbol{Y} = (Y_1, Y_2, \ldots, Y_P)',$$

where, for each l, $Y_l = X_{ij}$ for some i, j. We then construct n estimates of each Y_l from the bispectral and spectral estimates at the n points in the "fine" grid, $\{\omega_{i_p}, \omega_{j_q}\}$, the estimates of Y being asymptotically normally distributed (Brillinger, 1965). This provides us with n estimates of $\boldsymbol{Y}$, which we denote by $\boldsymbol{Y}_1, \boldsymbol{Y}_2, \ldots, \boldsymbol{Y}_n$. Under the null-hypothesis that the observed process is linear, the mean vector of the $\{\boldsymbol{Y}_i\}$ contains identical elements. The test for linearity based on the X_{ij} thus corresponds to the well-known test for "symmetry" in multivariate analysis (see, for example, Anderson, 1958), and proceeds as follows. Let

$$\bar{\boldsymbol{Y}} = \frac{1}{n}\sum_{i=1}^{n} \boldsymbol{Y}_i, \qquad \hat{\boldsymbol{S}}_Y = \sum_{i=1}^{n} (\boldsymbol{Y}_i - \bar{\boldsymbol{Y}})(\boldsymbol{Y}_i - \bar{\boldsymbol{Y}})', \qquad \hat{\boldsymbol{\Sigma}}_Y = \frac{1}{n}\hat{\boldsymbol{S}}_Y.$$

Define a column vector $\boldsymbol{\beta}$ of order $Q \times 1$, where $Q = P-1$, such that $\boldsymbol{\beta} = \boldsymbol{BY}$, where the $Q \times P$ matrix $\boldsymbol{B}$ takes the form,

$$\boldsymbol{B} = \begin{bmatrix} 1 & -1 & 0 & \cdots & 0 \\ 0 & 1 & -1 & \cdots & 0 \\ \cdots & \cdots & \cdots & \cdots & \cdots \\ 0 & 0 & 0 & 1 & -1 \end{bmatrix}.$$

Under the null hypotheses, $\boldsymbol{\beta}$ is asymptotically multivariate normal, with mean vector $\boldsymbol{O}$, and variance-covariance matrix $\boldsymbol{B\Sigma}_Y\boldsymbol{B}'$. The likelihood ratio test leads to the rejection of the null hypothesis if

$$T^2 = n\bar{\boldsymbol{\beta}}'\hat{\boldsymbol{S}}^{-1}\bar{\boldsymbol{\beta}} > \lambda_0,$$

where $\bar{\boldsymbol{\beta}} = \boldsymbol{B}\bar{\boldsymbol{Y}}$, $\hat{\boldsymbol{S}} = \boldsymbol{B}\hat{\boldsymbol{S}}_Y\boldsymbol{B}'$, and λ_0 is a constant determined by the significance level of the test. Under the null-hypothesis, the statistic

$$\mathscr{F}_2 = [(n-Q)/Q]T^2$$

has an F-distribution with $(Q, n-Q)$ degrees of freedom.

Choice of parameters

The first step is to choose the K equally spaced points in the interval $(0, \pi)$ which form the "course" grid, noting that K must be substantially smaller than N (the total number of observations in the series) in order to provide sufficient "space" around each pair of frequencies (ω_i, ω_j) to enable the construction of estimated spectra and bispectra at the neighbouring points $(\omega_{i_p}, \omega_{j_q})$ on the "fine" grid. The total number of points in each "fine" grid is $4r+1$, and, since there are $K^2/3$ grids, we require at least that $(4r+1)K^2/3 < N$. Also, to ensure that the spectral and bispectral estimates at different points on the "fine" grid are effectively uncorrelated, we require d to be chosen so that $\pi d/N$ is larger than the bandwidth of the spectral window used in the estimation procedure, in addition to which r must be chosen so that $h_3(\omega_1, \omega_2)$ and $h(\omega)$ are roughly constant over points in each "fine" grid—which implies that $2\pi r/N$ is less than the "bandwidths" of both h_3 and h. Finally, to ensure that points in different "fine" grids do not overlap, we require $d < N/\{K(r+1)\}$.

To implement the above restrictions we should, ideally, have *a priori* information on the spectral and bispectral bandwidths. In practice, such information would rarely be available in any precise form, and the choice of parameters would then be a matter of discretion, using whatever information one could gather on the "smoothness" of the spectra and bispectra. This situation is common to almost all types of spectral analysis, and is certainly not unique to the tests described above. Subba Rao and Gabr (1980, 1984) provide some useful guidelines on the choice of parameters, and suggest some typical values of K, L, P, r, and n, which can be used in practical applications.

It should be emphasized that the procedures described above do not provide complete tests for either Gaussianity or linearity. It is quite possible for a non-Gaussian process to have a zero bispectrum, and an exhaustive test would entail the testing of *all* the higher-order spectra. Indeed, it can be seen from (3.4.2) that the bispectrum $h_3(\omega_1, \omega_2)$, will vanish for a linear model provided only that $\beta = 0$, i.e. that e_t has zero third-order moment. Similarly, the constancy of $X(\omega_1, \omega_2)$ does not necessarily imply that the process conforms to a linear model. However, it is reasonable to suggest that in most practical situations deviations from Gaussianity or linearity would show up in the bispectrum, and this is the basis of the above tests.

Hinnich (1982) has proposed a modified version of the Subba Rao–Gabr tests. In the Subba Rao–Gabr approach the spectral and bispectral estimates on the "fine" frequency grid are treated simply as a "data set" from a multivariate normal distribution with an unknown variance-covariance matrix. Hinnich makes use of the known asymptotic expression for the variance–covariance matrix of the vector $\boldsymbol{\xi}$, and his tests are thus based on χ^2 rather than F-distributions.

3.4.1 Some Numerical Examples

Subba Rao and Gabr (1984) have applied the above tests to various simulated series generated from both linear and non-linear models, and have analysed also two well-known sets of real time series data, namely the sunspot series and the Canadian lynx series. Both these time series have been subjected to extensive analysis in recent years, and they share the common feature of exhibiting asymmetrical cyclical behaviour. The sunspot series studied here consists of 256 observations covering the years 1700–1959, and shows a cycle of approximately 11.3 years: the original data are given in Waldheimer (1961) and in Appendix 1. The Canadian lynx data refer to the number of lynx trapped per year in the MacKenzie River district of north-west Canada over the period 1821–1934, and consist of 114 observations: the data are given in Campbell and Walker (1977) and in Appendix 1. Several authors have suggested that it may be appropriate to make a logarithmic transformation on the Canadian lynx series in order to bring the data closer to normality, and by examining sample measures of skewness and kurtosis, Campbell and Walker (1977) conclude that the logarithmically transformed data are nearly Gaussian. It is therefore of interest to apply the above test for Gaussianity to both the original lynx series and the logarithmically transformed series. To provide estimates of the (second-order) spectra, Subba Rao and Gabr (1984) used a Daniell window, corresponding to the covariance lag window generator,

$$k(s) = \left(\frac{\sin \pi s}{\pi s}\right).$$

The bispectral estimates were computed using a (symmetric) two-dimensional form of the above window, as given in (3.2.23). Table 3.1 gives the value of the window parameter M (see Section 3.2.1), together with the values of the other parameters used in the construction of the test statistics.

In relation to the test for Gaussianity, the computed values of T^2, $\mathscr{F}_1$, and the upper 5% point of the F-distribution with (14, 4) degrees of freedom are given in Table 3.2.

Table 3.1

	N	M	K	L	d	r	P	n
Sunspot series	256	20	6	4	8	2	7	9
Canadian lynx series	114	16	6	4	3.5	2	7	9
Logarithm of Canadian lynx series	114	16	6	4	3.5	2	7	9

Table 3.2

	T^2	$\mathscr{F}_1$	Upper 5% point
Sunspot series	7531.7	2151.9	5.89
Canadian lynx series	3044.6	869.9	5.89
Logarithm of Canadian lynx series	1073.7	306.8	5.89

It is apparent that the values of the $\mathscr{F}_1$ statistic for all three series are massively significant, confirming the generally held view that these series possess a non-Gaussian structure. However, it will be observed that although the $\mathscr{F}_1$ statistic for the logarithmically transformed lynx series is considerably smaller than that of the original series, this transformation has not succeeded in producing a linear Gaussian form.

Proceeding now to the test for linearity, the values of T^2, $\mathscr{F}_2$, and the upper 5% point of the F-distribution with $(Q, n-Q) = (6, 3)$ degrees of freedom are as shown in Table 3.3.

These results indicate that both the sunspot series and the Canadian lynx series do not conform to a linear model, but the logarithm of the Canadian lynx series may be fitted by a linear, but non-Gaussian, model. (Similar conclusions were reported by Subba Rao and Gabr (1980) following their analyses of the same three series using a Parzen lag window instead of the Daniell window used for the above results.)

Table 3.3

	T^2	$\mathscr{F}_2$	Upper 5% point
Sunspot series	536.7	268.4	8.94
Canadian lynx series	1068.6	534.3	8.94
Logarithm of Canadian lynx series	9.7	4.9	8.94

Chapter 4

Some Special Non-linear Models

In Chapter 3 we discussed general forms of input/output relationships for non-linear systems. However, we recall that when dealing with linear systems we may often prefer to fit finite parameter ARMA-type models rather than use non-parametric methods to estimate transfer functions of general linear models of the form (1.4.10). Similarly, in studying non-linear systems, we may prefer to fit *finite parameter* models, provided, of course, that we can construct classes of such models which have an appeal to generality, i.e. which are capable of describing a large variety of non-linear structures. In this chapter we discuss some special models which are of current interest, and which have been successfully applied to a wide range of real data.

4.1 BILINEAR MODELS

Bilinear models were first introduced in the control theory literature—see for example, Ruberti *et al.* (1972), Mohler (1973), and Brockett (1976a).

In this context, interest centres on continuous time models which describe relationships between observable input and output processes of physical systems. In the case of time series modelling we are interested mainly in discrete time models, and the "input" here is white noise.

The general discrete time bilinear models takes the form,

$$X_t + \sum_{j=1}^{p} a_j X_{t-j} = \sum_{j=0}^{r} c_j e_{t-j} + \sum_{i=1}^{m} \sum_{j=1}^{k} b_{ij} X_{t-i} e_{t-j} \tag{4.1.1}$$

where $c_0 = 1$ and $\{e_t\}$ is a strict white noise process, i.e. a sequence of independent zero mean random variables. It is apparent that (4.1.1) is an extension of the (linear) ARMA model (2.2.6) obtained by adding a *bilinear form* in $\{X_{t-i}, e_{t-j}\}$ to the RHS of (2.2.6). In fact, if we set $b_{ij}=0$, all i, j, then (4.1.1) clearly reduces to (2.2.6), and thus the bilinear model includes as a special case the standard ARMA models. It is interesting to note that had we followed conventional statistical thinking we would have been tempted to generalize (2.2.6) by adding, say, quadratic, cubic, ... terms in $\{X_{t-j}\}$ to the LHS of (2.2.6), and similar non-linear terms in $\{e_{t-j}\}$ to the RHS of (2.2.6). However, it turns out that by adding a *bilinear* term to the RHS of (2.2.6) we obtain a more powerful and parsimonius non-linear model. In fact, Brockett (1976a) has shown that, with suitable choices of the model parameters, the bilinear model can approximate to an arbitrary degree of accuracy any "well-behaved" Volterra series relationship over a finite time interval. To some extent this parallels the corresponding property of ARMA models, namely that they can approximate to an arbitrary degree of accuracy any general linear relationship between $\{X_t\}$ and $\{e_t\}$. In this sense the bilinear models may be regarded as a "natural" non-linear extension of the ARMA model. However, it must be stressed that Brockett's result holds only over a *finite time interval*, and does not hold over an infinite time horizon. This limits somewhat the degree of generality of bilinear models, and it may be shown that certain non-linear characteristics (such as, for example, limit cycle behaviour) cannot be captured by bilinear models. However, bilinear models are certainly capable of exhibiting many other types of non-linear features, and to illustrate some typical non-linear characteristics we show in Figs 4.1, 4.2 and 4.3 realizations constructed by Subba Rao (1979) from the following bilinear models,

(i) $X_t = 0.4X_{t-1} + 0.4X_{t-1}e_{t-1} + e_t$

(ii) $X_t = 0.4X_{t-1} + 0.8X_{t-1}e_{t-1} + e_t$

(iii) $X_t = 0.8X_{t-1} - 0.4X_{t-2} + 0.6X_{t-1}e_{t-1} + 0.7X_{t-2}e_{t-1} + e_t.$

In each case the $\{e_t\}$ are independent $N(0, 1)$ variables, and each realization consists of 1000 data points. It will be noticed that series (i), for which the coefficient of the bilinear term is fairly small, has a more or less conventional form, whereas series (ii), with a larger bilinear coefficient, exhibits a number of "bursts" of large-amplitude excursions. Series (iii), with further bilinear terms, exhibits an extreme form of this effect. (Note that the scale of the vertical axis of Fig. 4.3 has been compressed to such an extent that the fluctuations in the series outside the main "burst" are no longer visible.) This type of behaviour is a well-known feature of certain types of seismological data; particularly in series relating to earthquakes and underground explosions (see, for example, Dargahi-Noubary *et al.*, 1978). It is clearly indicative of a *non-Gaussian structure*, and it suggests that suitable models

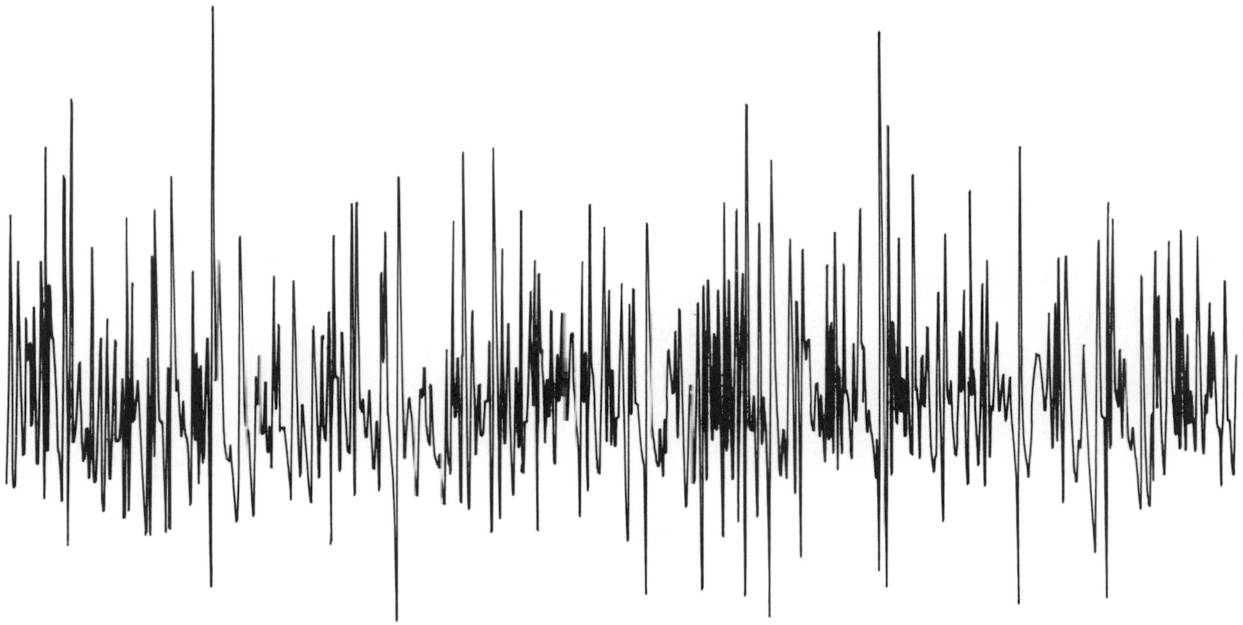

Fig. 4.1. $X_t = 0.4X_{t-1} + 0.4X_{t-1}e_{t-1} + e_t$.

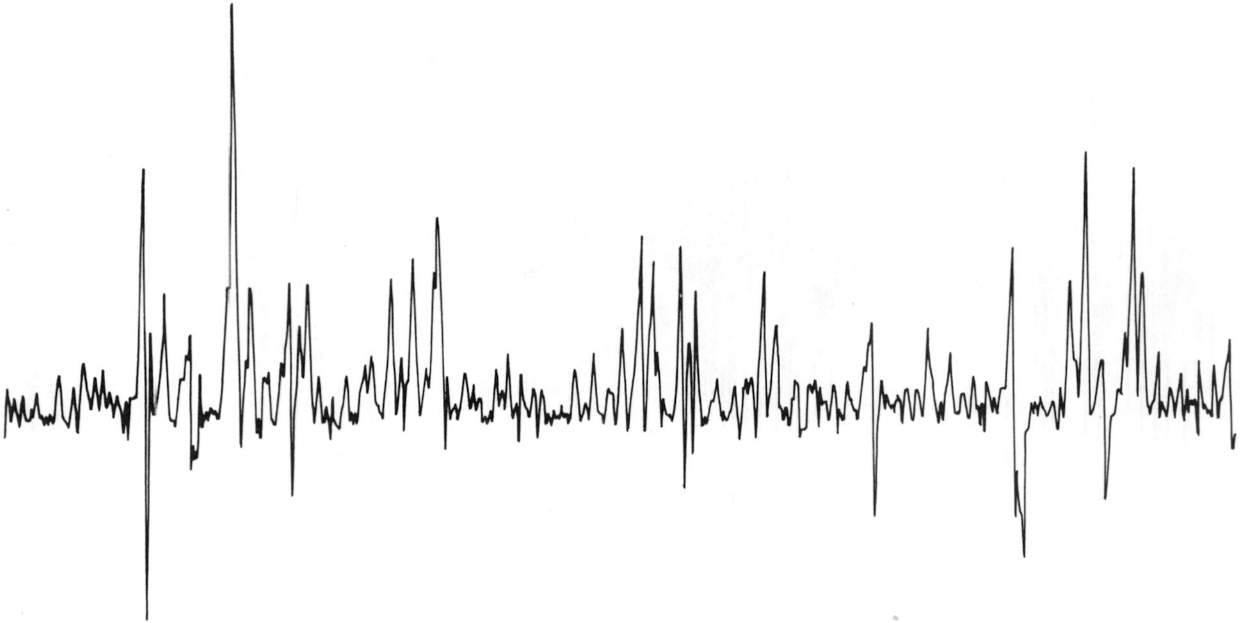

Fig. 4.2. $X_t = 0.4X_{t-1} + 0.8X_{t-1}e_{t-1} + e_t$.

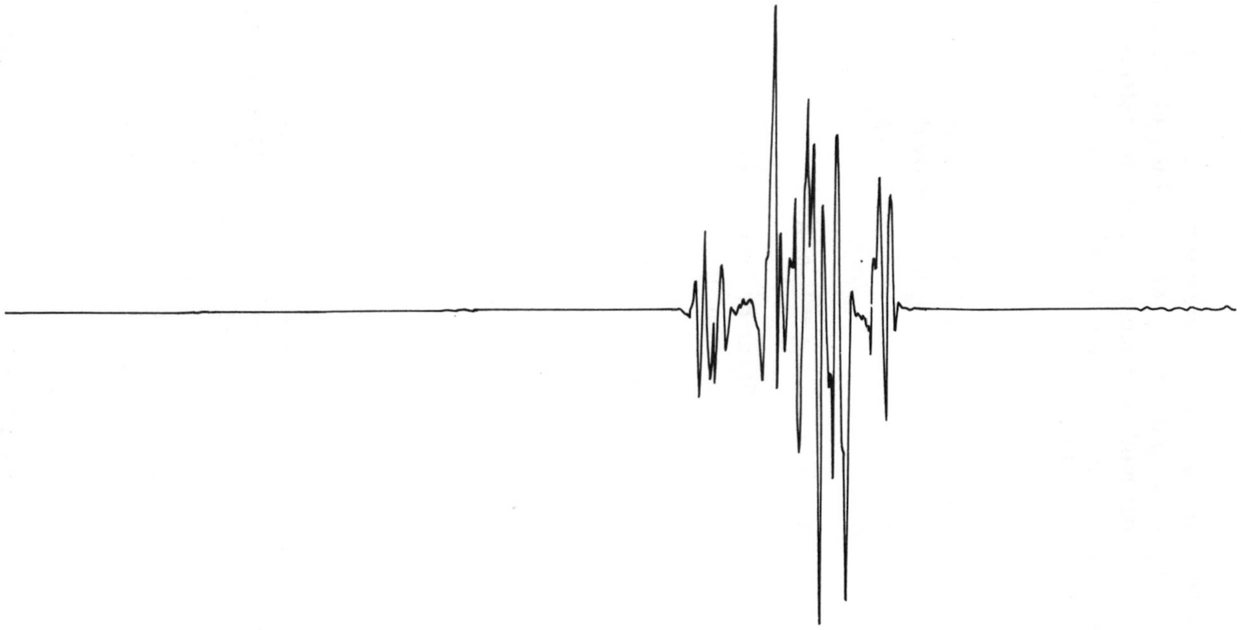

Fig. 4.3. $X_t = 0.8X_{t-1} - 0.4X_{t-2} + 0.6X_{t-1}e_{t-1} + 0.7X_{t-2}e_{t-1} + e_t$.

for such data would lead to marginal distributions for which certain moments fail to exist.

Bilinear models of the form (4.1.1) were first introduced into the statistical literature by Granger and Andersen (1978), who also studied the statistical properties of first-order models. A more general and systematic study of bilinear models was later given by Subba Rao (1981) and in the monograph by Subba Rao and Gabr (1984).

Vector form of bilinear model

Following Subba Rao (1981) we will denote the general bilinear models (4.1.1) by BL(p, r, m, k), the integers p, r, m, k clearly denoting the orders of the various terms in (4.1.1). Consider now the BL(p, 0, p, 1) model, namely

$$X_t + \sum_{j=1}^{p} a_j X_{t-j} = e_t + \sum_{i=1}^{p} b_{i1} X_{t-i} e_{t-1}. \tag{4.1.2}$$

Just as the ARMA model (2.2.6) can be rewritten in an equivalent vector state-space form (2.3.5), (2.3.6), we can similarly rewrite (4.1.2) in vector form by introducing the matrices,

$$\boldsymbol{A}_{p\times p} = \begin{bmatrix} -a_1 & -a_2 & \dots & -a_p \\ 1 & 0 & \dots & 0 \\ 0 & 1 & \dots & 0 \\ 0 & 0 & 1 & 0 \end{bmatrix}, \qquad \boldsymbol{B}_{p\times p} = \begin{bmatrix} b_{11} & b_{21} & \dots & b_{p1} \\ 0 & 0 & \dots & 0 \\ \dots & \dots & \dots & \dots \\ 0 & 0 & & 0 \end{bmatrix}$$

$$\boldsymbol{C}_{p\times 1} = (1, 0, \dots, 0)', \qquad \boldsymbol{H}_{1\times p} = (1, 0, \dots, 0),$$

and the "state-vector", $\boldsymbol{x}_t = (X_t, X_{t-1}, \dots, X_{t-p+1})'$. It is then easily verified that (4.1.2) can be rewritten as

$$\begin{aligned} \boldsymbol{x}_t &= \boldsymbol{A}\boldsymbol{x}_{t-1} + \boldsymbol{B}\boldsymbol{x}_{t-1}e_{t-1} + \boldsymbol{C}e_t \\ X_t &= \boldsymbol{H}\boldsymbol{x}_t. \end{aligned} \tag{4.1.3}$$

We will now denote (4.1.3) by VBL(p), the initial letter indicating that (4.1.3) is written in vector form. Although we have referred loosely to $\boldsymbol{x}_t$ as the "state-vector", it should be noted that (4.1.3) is *not* a Markovian representation. However, Tuan Pham Dinh and Lanh Tat Tran (1981) have shown that a Markovian representation can be derived for the vector $\boldsymbol{Z}_t = (\boldsymbol{A} + \boldsymbol{B}e_t)\boldsymbol{x}_t$. Specifically, we can rewrite (4.1.3) in the form

$$\begin{aligned} \boldsymbol{Z}_t &= (\boldsymbol{A} + \boldsymbol{B}e_t)\boldsymbol{Z}_{t-1} + (\boldsymbol{A} + \boldsymbol{B}e_t)\boldsymbol{C}e_t \\ \boldsymbol{x}_t &= \boldsymbol{Z}_{t-1} + \boldsymbol{C}e_t. \end{aligned}$$

Tuan Pham Dinh (1983) has used this Markovian representation to obtain the moments of $\boldsymbol{Z}_t$ in the case of first-order models, i.e. when $\boldsymbol{Z}_t$ is a scalar process.

It may be noted that (4.1.3) is the discrete-time analogue of the type of bilinear models most frequently studied in the control theory literature. The continuous time version of (4.1.3) is given by

$$\dot{\boldsymbol{x}}(t) = \boldsymbol{A}\boldsymbol{x}(t) + \boldsymbol{B}\boldsymbol{x}(t)e(t) + \boldsymbol{C}e(t). \tag{4.1.3a}$$

Consider now the BL(p, 0, p, k) model, namely

$$X_t + \sum_{j=1}^{p} a_j X_{t-j} = e_t + \sum_{i=1}^{p}\sum_{j=1}^{k} b_{ij} X_{t-i} e_{t-j}. \tag{4.1.4}$$

The vector representation is readily extended to this case as follows: let $\boldsymbol{A}$, $\boldsymbol{C}$, $\boldsymbol{H}$, $\boldsymbol{x}_t$ be defined as above, and let the $p \times p$ matrix $\boldsymbol{B}_j$ be given by

$$\boldsymbol{B}_j = \begin{bmatrix} b_{1j} & b_{2j} & \cdots & b_{pj} \\ 0 & 0 & \cdots & 0 \\ \cdots & \cdots & \cdots & \cdots \\ 0 & 0 & \cdots & 0 \end{bmatrix}, \qquad j = 1, \ldots, k.$$

Then again it is easily verified that (4.1.4) may be rewritten as

$$\begin{aligned} \boldsymbol{x}_t &= \boldsymbol{A}\boldsymbol{x}_{t-1} + \sum_{j=1}^{k} \boldsymbol{B}_j \boldsymbol{x}_{t-1} e_{t-j} + \boldsymbol{C}e_t \\ X_t &= \boldsymbol{H}\boldsymbol{x}_t, \end{aligned} \tag{4.1.5}$$

and we will refer to the above as the VBL(p, k) representation.

Finally, we note that the BL(p, r, m, k) model can be written as

$$\begin{aligned} \boldsymbol{x}_t &= \boldsymbol{A}\boldsymbol{x}_{t-1} + \sum_{j=1}^{k} \boldsymbol{B}_j \boldsymbol{x}_{t-1} e_{t-j} + \boldsymbol{C}e_t \\ X_t &= \boldsymbol{H}\boldsymbol{x}_t \end{aligned} \tag{4.1.6}$$

where, writing $p' = \max(p, m)$, $\boldsymbol{x}_t$ now has dimensions p',

$$\boldsymbol{A}_{p'\times p'} = \begin{bmatrix} -a_1 & -a_2 & \cdots & -a_{p'} \\ 1 & 0 & \cdots & 0 \\ 0 & 1 & \cdots & 0 \\ \cdots & \cdots & \cdots & \cdots \\ 0 & 0 & 1 & 0 \end{bmatrix}, \qquad \boldsymbol{C}_{p'\times r} = \begin{bmatrix} c_0 & c_1 & \cdots & c_r \\ 0 & 0 & \cdots & 0 \\ \cdots & \cdots & \cdots & \cdots \\ 0 & 0 & \cdots & 0 \end{bmatrix},$$

the $p' \times p'$ matrix $\boldsymbol{B}_j$ is given by

$$\boldsymbol{B}_j = \begin{bmatrix} b_{1j} & \cdots & b_{p'j} \\ 0 & \cdots & 0 \\ 0 & \cdots & 0 \end{bmatrix}, \qquad j = 1, \ldots, k,$$

and $\mathbf{e}_t = (e_t, e_{t-1}, \ldots, e_{t-r})$. (In the above, $a_i = 0$, $i > p$, and $b_{ij} = 0$, $i > m$.)

Volterra series expansions for bilinear models

A bilinear model describes a non-linear relationship between $\{X_t\}$ and $\{e_t\}$, and in principle the relationship can also be described by a Volterra series expansion. Priestley (1978b) considered the first-order model, BL(1, 0, 1, 1).

$$X_t + aX_{t-1} = e_t + bX_{t-1}e_{t-1} \tag{4.1.7}$$

and studied the form of the generalized transfer functions for this case. Setting $e_t = e^{i\omega t}$ in (4.1.7) and writing

$$X_t = \Gamma_1(\omega)e^{i\omega t} + \Gamma_2(\omega, \omega)e^{2i\omega t} + \cdots \tag{4.1.8}$$

(cf. (3.1.7)), we obtain on comparing the coefficients of $e^{i\omega t}$ on both sides of (4.1.7),

$$(1 + ae^{-i\omega})\Gamma_1(\omega) = 1,$$

i.e.

$$\Gamma_1(\omega) = \frac{1}{(1 + ae^{-i\omega})}. \tag{4.1.9}$$

Similarly, setting $e_t = e^{i\omega_1 t} + e^{i\omega_2 t}$, and writing

$$\begin{aligned} X_t = {} & \Gamma_1(\omega_1)\, e^{i\omega_1 t} + \Gamma_1(\omega_2)\, e^{i\omega_2 t} + \Gamma_2(\omega_1, \omega_1)\, e^{2i\omega_1 t} + \Gamma_2(\omega_2, \omega_2)\, e^{2i\omega_2 t} \\ & + \{\Gamma_2(\omega_1, \omega_2) + \Gamma_2(\omega_2, \omega_1)\}\, e^{i(\omega_1 + \omega_2)t} + \cdots, \end{aligned} \tag{4.1.10}$$

we find on comparing coefficients of $e^{i(\omega_1+\omega_2)t}$ on both sides (and assuming w.l.o.g., that $\Gamma_2(\omega_1, \omega_2)$ is symmetric in ω_1, ω_2),

$$2(1 + a\, e^{-i(\omega_1+\omega_2)})\Gamma_2(\omega_1, \omega_2) = b\{\Gamma_1(\omega_1) + \Gamma_1(\omega_2)\}\, e^{-i(\omega_1+\omega_2)},$$

or

$$\begin{aligned} \Gamma_2(\omega_1, \omega_2) &= \frac{b}{2(e^{i(\omega_1+\omega_2)} + a)}\{\Gamma_1(\omega_1) + \Gamma_1(\omega_2)\} \\ &= \frac{b}{2(e^{i(\omega_1+\omega_2)} + a)}\left\{\frac{1}{1 + ae^{-i\omega_1}} + \frac{1}{1 + ae^{-i\omega_2}}\right\}. \end{aligned} \tag{4.1.11}$$

In particular, along the diagonal $\omega_1 = \omega_2 = \omega$,

$$\Gamma_2(\omega, \omega) = \frac{b\, e^{i\omega}}{(e^{2i\omega} + a)(e^{i\omega} + a)}, \tag{4.1.12}$$

and a similar analysis gives for the kth-order transfer function,

$$\Gamma_k(\omega, \omega, \ldots, \omega) = \frac{b^{k-1} e^{i\omega}}{(e^{ik\omega} + a)(e^{i(k-1)\omega} + a)\ldots(e^{i\omega} + a)}. \tag{4.1.13}$$

It is interesting to observe that the function $\Gamma_1(\omega)$ is identical with the conventional transfer function of the linear model obtained by setting $b = 0$ in (4.1.7), and thus it contains no information on the non-linear component of the model. However, the quadratic transfer function, $\Gamma_2(\omega_1, \omega_2)$, involves both the parameters a and b, and thus the model may be determined from a knowledge of this function. In fact, all we need to know is the form of $\Gamma_2(\omega_1, \omega_2)$ along the diagonal; the function $\Gamma_2(\omega, \omega)$ determines both parameters and thus determines all the higher-order transfer functions for this model.

Subba Rao and Gabr (1984) have extended the above analysis, and have shown that for the VBL(p) model (4.1.3),

$$\Gamma_1(\omega) = \boldsymbol{H}(\boldsymbol{I} - \boldsymbol{A}\, e^{-i\omega})^{-1}\boldsymbol{C} \tag{4.1.14}$$

$$\begin{aligned}\Gamma_2(\omega_1, \omega_2) = \tfrac{1}{2}[&\boldsymbol{H}(\boldsymbol{I} - \boldsymbol{A}\, e^{-i(\omega_1+\omega_2)})^{-1}\boldsymbol{B}(\boldsymbol{I} - \boldsymbol{A}\, e^{i\omega_1})^{-1}\boldsymbol{C} \\ &+ \boldsymbol{H}(\boldsymbol{I} - \boldsymbol{A}\, e^{-i(\omega_1+\omega_2)})^{-1}\boldsymbol{B}(\boldsymbol{I} - \boldsymbol{A}e^{-i\omega_2})^{-1}\boldsymbol{C}]\, e^{-i(\omega_1+\omega_2)}.\end{aligned} \tag{4.1.15}$$

These expressions reduce to (4.1.9), (4.1.11) for the BL(1, 0, 1, 1) model.

Asymptotic stationarity and covariances

We now turn to the BL(p, 0, p, 1) model, or rather its vector form VBL(p) given by (4.1.3), and obtain conditions for asymptotic stationarity and expressions for the covariances. We follow the approach given by Subba Rao (1981), which is based on the assumption that the $\{e_t\}$ are i.i.d. $N(0, 1)$ variables.

From (4.1.3) we have,

$$E(X_t) = \boldsymbol{H}E(\boldsymbol{x}_t)$$

$$\operatorname{cov}(X_t, X_{t+s}) = \boldsymbol{H}[E\{(\boldsymbol{x}_t - E(\boldsymbol{x}_t)(\boldsymbol{x}_{t+s} - E(\boldsymbol{x}_{t+s}))'\}]\boldsymbol{H}',$$

and thus it suffices to consider the second-order properties of the vector process $\boldsymbol{x}_t$.

Write $\boldsymbol{\mu}_t = E(\boldsymbol{x}_t)$, $\boldsymbol{V}_t = E(\boldsymbol{x}_t\boldsymbol{x}_t')$

$$\boldsymbol{S}_t = E(\boldsymbol{x}_t\boldsymbol{x}_t' e_t), \qquad \boldsymbol{W}_t = E(\boldsymbol{x}_t\boldsymbol{x}_t' e_t^2).$$

Taking expectations of both sides of (4.1.3) (with t replaced by $(t+1)$), and using $E(\boldsymbol{x}_t e_t) = \boldsymbol{C}$, we obtain

$$\boldsymbol{\mu}_{t+1} = \boldsymbol{A}\boldsymbol{\mu}_t + \boldsymbol{B}\boldsymbol{C},$$

which, on repeated application, gives

$$\boldsymbol{\mu}_{t+1} = \boldsymbol{A}^t\boldsymbol{\mu}_1 + \left[\sum_{j=0}^{t-1} \boldsymbol{A}^j\right] \boldsymbol{B}\boldsymbol{C}. \tag{4.1.16}$$

If $\boldsymbol{B} = 0$, and $\boldsymbol{\mu}_1 = 0$, then $\boldsymbol{\mu}_t = 0$, all $t \geq 1$, and the process is then stationary to order 1 without any conditions on $\boldsymbol{A}$. More generally, we may derive the condition for first-order stationarity as follows. Let $\rho(\boldsymbol{A})$ denote the spectral radius of the matrix $\boldsymbol{A}$, i.e.

$$\rho(\boldsymbol{A}) = \max_i |\lambda_i(\boldsymbol{A})| \tag{4.1.17}$$

where $\{\lambda_i(\boldsymbol{A})\}$ are the eigenvalues of $\boldsymbol{A}$. It is then well known that a sufficient condition for the limit of the RHS of (4.1.16) to be finite at $t \to \infty$ is that

$$\rho(\boldsymbol{A}) < 1.$$

When this condition holds the process is (asymptotically) stationary to order 1, and the limiting mean $\boldsymbol{\mu}$ is given by

$$\boldsymbol{\mu} = (\boldsymbol{I} - \boldsymbol{A})^{-1}\boldsymbol{B}\boldsymbol{C}. \tag{4.1.18}$$

Turning now to the second-order moments, we first note that

$$E(\boldsymbol{x}_t e_{t+1}) = E(\boldsymbol{x}_t e_t e_{t+1}) = 0.$$

Substituting the expression for $\boldsymbol{x}_t$ given by (4.1.3) into the definition of $\boldsymbol{V}_t$ we then obtain,

$$\boldsymbol{V}_t = \boldsymbol{A}\boldsymbol{V}_{t-1}\boldsymbol{A}' + \boldsymbol{A}\boldsymbol{S}_{t-1}\boldsymbol{B}' + \boldsymbol{B}\boldsymbol{S}_{t-1}\boldsymbol{A}' + \boldsymbol{B}\boldsymbol{W}_{t-1}\boldsymbol{B}' + \boldsymbol{C}\boldsymbol{C}'. \tag{4.1.19}$$

A similar substitution into the definition of $\boldsymbol{S}_t$ gives (recalling that $E(e_t^2) = 1$, $E(\boldsymbol{x}_t e_t) = \boldsymbol{C}$),

$$\boldsymbol{S}_t = \boldsymbol{A}\boldsymbol{\mu}_{t-1}\boldsymbol{C}' + \boldsymbol{B}\boldsymbol{C}\boldsymbol{C}' + \boldsymbol{C}\boldsymbol{\mu}'_{t-1}\boldsymbol{A}' + \boldsymbol{C}\boldsymbol{C}'\boldsymbol{B}', \tag{4.1.20}$$

and the same operation for $\boldsymbol{W}_t$ yields (recalling that since the e_t are $N(0, 1)$, $E(e_t^4) = 3$),

$$\boldsymbol{W}_t = \boldsymbol{A}\boldsymbol{V}_{t-1}\boldsymbol{A}' + \boldsymbol{A}\boldsymbol{S}_{t-1}\boldsymbol{B}' + \boldsymbol{B}\boldsymbol{S}_{t-1}\boldsymbol{A}' + \boldsymbol{B}\boldsymbol{W}_{t-1}\boldsymbol{B}' + 3\boldsymbol{C}\boldsymbol{C}'$$

or

$$\boldsymbol{W}_t = \boldsymbol{V}_t + 2\boldsymbol{C}'\boldsymbol{C}. \tag{4.1.21}$$

Equations (4.1.19), (4.1.21) now give

$$\boldsymbol{V}_t = \boldsymbol{A}\boldsymbol{V}_{t-1}\boldsymbol{A}' + \boldsymbol{A}\boldsymbol{S}_{t-1}\boldsymbol{B}' + \boldsymbol{B}\boldsymbol{S}_{t-1}\boldsymbol{A}' + \boldsymbol{B}\boldsymbol{V}_{t-1}\boldsymbol{B}' + 2\boldsymbol{B}\boldsymbol{C}\boldsymbol{C}'\boldsymbol{B}' + \boldsymbol{C}\boldsymbol{C}'. \tag{4.1.22}$$

Assuming that the process is stationary to order 1, and that t is sufficiently large, we may set $\boldsymbol{\mu}_t = \boldsymbol{\mu}$, which in turn implies that $\boldsymbol{S}_t = \boldsymbol{S}$, where

$$\boldsymbol{S} = \boldsymbol{A}\boldsymbol{\mu}\boldsymbol{C}' + \boldsymbol{B}\boldsymbol{C}\boldsymbol{C}' + \boldsymbol{C}\boldsymbol{\mu}'\boldsymbol{A}' + \boldsymbol{C}\boldsymbol{C}'\boldsymbol{B}'.$$

Equation (4.1.22) can now be written as

$$\boldsymbol{V}_t = \boldsymbol{A}\boldsymbol{V}_{t-1}\boldsymbol{A}' + \boldsymbol{B}\boldsymbol{V}_{t-1}\boldsymbol{B}' + \boldsymbol{\Delta}_1 \tag{4.1.23}$$

where

$$\boldsymbol{\Delta}_1 = \boldsymbol{A}\boldsymbol{S}\boldsymbol{B}' + \boldsymbol{B}\boldsymbol{S}\boldsymbol{A}' + 2\boldsymbol{B}\boldsymbol{C}\boldsymbol{C}'\boldsymbol{B}' + \boldsymbol{C}\boldsymbol{C}'.$$

To investigate the conditions under which $\boldsymbol{V}_t$ attains a limiting value $\boldsymbol{V}$ as $t \to \infty$, we now write (4.1.23) in the form

$$\text{vec}(\boldsymbol{V}_t) = [\boldsymbol{A}\otimes\boldsymbol{A} + \boldsymbol{B}\otimes\boldsymbol{B}]\,\text{vec}(\boldsymbol{V}_{t-1}) + \text{vec}\,\boldsymbol{\Delta}_1, \tag{4.1.24}$$

where, for any matrix $\boldsymbol{M}$ of order $p \times p$ say, $\text{vec}(\boldsymbol{M})$ denotes the $p^2 \times 1$ vector obtained by setting down the column of $\boldsymbol{M}$ underneath each other, and, for example, $\boldsymbol{A}\otimes\boldsymbol{A}$ denotes the Kronecker product of $\boldsymbol{A}$ with itself (a matrix of order $p^2 \times p^2$). Equation (4.1.24) now gives us a first-order difference equation for $\text{vec}(\boldsymbol{V}_t)$, and the solution can be derived as a power series in $[\boldsymbol{A}\otimes\boldsymbol{A} + \boldsymbol{B}\otimes\boldsymbol{B}]$. It then follows that a sufficient condition for the solution for $\text{vec}(\boldsymbol{V}_t)$ to converge to a limiting value as $t \to \infty$ is that

$$\rho[\boldsymbol{A}\otimes\boldsymbol{A} + \boldsymbol{B}\otimes\boldsymbol{B}] < 1. \tag{4.1.25}$$

The above is thus a sufficient condition for the process $\boldsymbol{x}_t$ to be asymptotically stationary to order 2.

Note that in deriving the above results we have assumed that $E(e_t^2) = 1$. If $E(e_t^2) = \sigma_e^2 \neq 1$, then simple modification of the above analysis shows that (4.1.25) should be replaced by

$$\rho[\boldsymbol{A}\otimes\boldsymbol{A} + \boldsymbol{B}\otimes\boldsymbol{B}\sigma_e^2] < 1.$$

The condition (4.1.25) is a somewhat weaker form of the stationarity condition originally derived by Subba Rao (1979), namely

$$\|\boldsymbol{A}\|^2 + \|\boldsymbol{B}\|^2 < 1, \tag{4.1.25a}$$

where $\|\boldsymbol{A}\|^2 =$ largest eigenvalue of $\boldsymbol{A}\boldsymbol{A}'$, and $\|\boldsymbol{B}\|^2$ is similarly defined. To see the relationship between (4.1.25) and (4.1.25a) we first recall the well-known result that (Wilkinson, 1965, p. 59)

$$\rho(\boldsymbol{A}) \leq \|\boldsymbol{A}\|,$$

with equality holding if $\boldsymbol{A}$ is a symmetric matrix. (If $\boldsymbol{A}$ is symmetric then its eigenvalues, $\lambda_1(\boldsymbol{A}), \ldots, \lambda_p(\boldsymbol{A})$ are real, and the eigenvalues of $\boldsymbol{A}\boldsymbol{A}'$ are

$\lambda_1^2(\boldsymbol{A}), \ldots, \lambda_p^2(\boldsymbol{A})$. Suppose the eigenvalues are labelled so that $\lambda_1^2(\boldsymbol{A}) < \cdots < \lambda_p^2(\boldsymbol{A})$. Then in this case, $\|\boldsymbol{A}\| = \{\lambda_p^2(\boldsymbol{A})\}^{1/2} = |\lambda_p(\boldsymbol{A})| = \rho(\boldsymbol{A})$.)

We then have,

$$\rho(\boldsymbol{A}\otimes\boldsymbol{A}+\boldsymbol{B}\otimes\boldsymbol{B}) \le \|\boldsymbol{A}\otimes\boldsymbol{A}+\boldsymbol{B}\otimes\boldsymbol{B}\|$$
$$\le \|\boldsymbol{A}\otimes\boldsymbol{A}\| + \|\boldsymbol{B}\otimes\boldsymbol{B}\|,$$

using the triangular inequality for matrix norms. Also, we have the following results for Kronecker products of suitably dimensioned matrices (see, for example, Householder, 1964, p. 35):

(i) $(\boldsymbol{A}\otimes\boldsymbol{B})' = \boldsymbol{A}'\otimes\boldsymbol{B}'$
(ii) $(\boldsymbol{A}_1\otimes\boldsymbol{A}_2)(\boldsymbol{B}_1\otimes\boldsymbol{B}_2) = \boldsymbol{A}_1\boldsymbol{B}_1\otimes\boldsymbol{A}_2\boldsymbol{B}_2$
(iii) the eigenvalues of $\boldsymbol{A}\otimes\boldsymbol{B}$ are $\{\lambda_i(\boldsymbol{A})\lambda_j(\boldsymbol{B})\}$, $i, j = 1, 2, \ldots, p$.

It then follows that

$$(\boldsymbol{A}\otimes\boldsymbol{A})(\boldsymbol{A}\otimes\boldsymbol{A})' = (\boldsymbol{A}\otimes\boldsymbol{A})(\boldsymbol{A}'\otimes\boldsymbol{A}') = \boldsymbol{A}\boldsymbol{A}'\otimes\boldsymbol{A}\boldsymbol{A}',$$

so that

$$\|\boldsymbol{A}\otimes\boldsymbol{A}\| = [\text{largest eigenvalue of } (\boldsymbol{A}\boldsymbol{A}')\otimes(\boldsymbol{A}\boldsymbol{A}')]^{1/2}.$$

But $\boldsymbol{A}\boldsymbol{A}'$ is a symmetric positive semi-definite matrix, and hence its eigenvalues are real and non-negative. Let these eigenvalues be denoted by

$$0 \le \delta_1(\boldsymbol{A}\boldsymbol{A}') < \cdots < \delta_p(\boldsymbol{A}\boldsymbol{A}').$$

Then,

$$\|\boldsymbol{A}\otimes\boldsymbol{A}\| = [\max_{i,j}\{\delta_i(\boldsymbol{A}\boldsymbol{A}')\delta_j(\boldsymbol{A}\boldsymbol{A}')\}]^{1/2}$$
$$= \delta_p(\boldsymbol{A}\boldsymbol{A}')$$
$$= \|\boldsymbol{A}\|^2.$$

Using the same result for $\|\boldsymbol{B}\otimes\boldsymbol{B}\|$ we finally obtain

$$\rho(\boldsymbol{A}\otimes\boldsymbol{A}+\boldsymbol{B}\otimes\boldsymbol{B}) \le \|\boldsymbol{A}\otimes\boldsymbol{A}\| + \|\boldsymbol{B}\otimes\boldsymbol{B}\|$$
$$= \|\boldsymbol{A}\|^2 + \|\boldsymbol{B}\|^2.$$

Hence condition (4.1.25a) implies condition (4.1.25).

Assuming that the condition (4.1.25) holds, and that the process has attained its stationary state, we can then obtain an expression for the stationary variance–covariance matrix of $\boldsymbol{x}_t$ as follows. Writing $\boldsymbol{V} = E(\boldsymbol{x}_t\boldsymbol{x}_t')$ we have from the stationary form of (4.1.23),

$$\boldsymbol{V} = \boldsymbol{A}\boldsymbol{V}\boldsymbol{A}' + \boldsymbol{B}\boldsymbol{V}\boldsymbol{B}' + \boldsymbol{\Delta}_1, \tag{4.1.26}$$

and this equation, being linear in V, can be solved explicitly. Again assuming stationarity, the higher-order covariance matrices can be obtained directly from (4.1.3). Multiplying both sides of (4.1.3) (with t replaced by $(t+1)$) by x_t' and taking expectations gives

$$E(\boldsymbol{x}_{t+1}\boldsymbol{x}_t') = \boldsymbol{A}E(\boldsymbol{x}_t\boldsymbol{x}_t') + \boldsymbol{B}E(\boldsymbol{x}_t\boldsymbol{x}_t'e_t). \qquad (4.1.27)$$

Similarly, using (4.1.3) to express $\boldsymbol{x}_{t+s}$ in terms of $\boldsymbol{x}_{t+1}$, multiplying by $\boldsymbol{x}_t'$, and taking expectations gives for $s \geq 2$,

$$E(\boldsymbol{x}_{t+s}\boldsymbol{x}_t') = \boldsymbol{A}^{s-1}E(\boldsymbol{x}_{t+1}\boldsymbol{x}_t') + \left\{\sum_{j=0}^{s-2} \boldsymbol{A}^j\boldsymbol{BC}\right\}\boldsymbol{\mu}'. \qquad (4.1.28)$$

Now let

$$\boldsymbol{C}(s) = E[(\boldsymbol{x}_{t+s} - \boldsymbol{\mu})(\boldsymbol{x}_t - \boldsymbol{\mu})']$$

denote the autocovariance matrix of lag s for $\boldsymbol{x}_t$. Using the results (4.1.26) and (4.1.27) we now have,

$$\boldsymbol{C}(0) = \boldsymbol{AC}(0)\boldsymbol{A}' + \boldsymbol{BC}(0)\boldsymbol{B}' + \boldsymbol{\Delta}_2, \qquad (4.1.29)$$

$$\boldsymbol{C}(1) = \boldsymbol{AC}(0) + \boldsymbol{\Delta}_3 \qquad (4.1.30)$$

$$\boldsymbol{C}(s) = \boldsymbol{AC}(s-1) = \boldsymbol{A}^{s-1}\boldsymbol{C}(1), \qquad s = 2, 3, \ldots. \qquad (4.1.31)$$

where

$$\boldsymbol{\Delta}_2 = \boldsymbol{B\mu\mu}'\boldsymbol{B}' + \boldsymbol{A\mu\mu}'\boldsymbol{A}' + \boldsymbol{ASB}' + \boldsymbol{BS}'\boldsymbol{A}' + 2\boldsymbol{BCC}'\boldsymbol{B}' + \boldsymbol{CC}' - \boldsymbol{\mu\mu}'$$

$$\boldsymbol{\Delta}_3 = \boldsymbol{A\mu\mu}' + \boldsymbol{BS} - \boldsymbol{\mu\mu}'.$$

If we now use the explicit form of the matrices $\boldsymbol{A}$, $\boldsymbol{B}$, which were used to derive (4.1.3) from (4.1.2), we obtain from (4.1.31),

$$R(s) + a_1R(s-1) + \cdots + a_pR(s-p) = 0, \qquad s \geq 2 \qquad (4.1.32)$$

where $R(s) = \text{cov}(X_{t+s}, X_t)$ denotes the (scalar) autocovariance function of the original process $\{X_t\}$. The striking feature of (4.1.32) is that *it has exactly the same form as the Yule–Walker equations for the autocovariance function of an ARMA$(p, 1)$ model.* This result thus establishes that a BL$(p, 0, p, 1)$ model has essentially the same covariance structure as an ARMA$(p, 1)$ model, and it highlights the fact that in order to distinguish between linear and non-linear models we need to study moments higher than the second order.

Homogeneous bilinear models

If $\boldsymbol{C} = 0$ in (4.1.3) the model is called *homogeneous*, and we note that in this case there is no e_{t+1} term present, i.e. there is no independent "shock" arriving at time $(t+1)$. In this case we have $\boldsymbol{\mu}=\mathbf{0}$, $\boldsymbol{S}=\mathbf{0}$, and hence (4.1.23) reduces to

$$\boldsymbol{V}_t=\boldsymbol{A}\boldsymbol{V}_{t-1}\boldsymbol{A}'+\boldsymbol{B}\boldsymbol{V}_{t-1}\boldsymbol{B}'.$$

If initially we have $\boldsymbol{V}_1=\mathbf{0}$, then clearly $\boldsymbol{V}_t=\mathbf{0}$, all $t\geq 1$. More generally, the discussion following (4.1.23) shows that a homogeneous model degenerates into a deterministic form as $t\to\infty$ if $\rho[\boldsymbol{A}\otimes\boldsymbol{A}+\boldsymbol{B}\otimes\boldsymbol{B}\}<1$, but if $\rho\{\boldsymbol{A}\otimes\boldsymbol{A}+\boldsymbol{B}\otimes\boldsymbol{B}\}>1$, the model "explodes" as $t\to\infty$.

First-order model

It is instructive to apply the above results to the simple case of the first order BL(1, 0, 1, 1) model, namely

$$X_t+aX_{t-1}=e_t+bX_{t-1}e_{t-1}. \tag{4.1.33}$$

Here, the condition (4.1.25) for asymptotic stationarity becomes

$$a^2+b^2<1,$$

and (4.1.18), (4.1.29), (4.1.30) give the following

$$E(\boldsymbol{X}_t)\sim\frac{b}{1+a}$$

$$E(X_t^2)\sim\frac{1+2b^2}{1-a^2-b^2}-\frac{4ab^2}{(1+a)(1-a^2-b^2)}$$

$$E(X_{t+1}X_t)\sim -aE(X_t^2)+\frac{2b^2}{(1+a)}.$$

Some further special types of bilinear models

The preceding discussion has concentrated mainly on the BL(p, 0, p, 1) model (and its vector counterpart, the VBL(p) model). Two other special forms of the general model (4.1.1) are of interest, namely the *diagonal model*,

$$X_t+\sum_{i=1}^{l}a_iX_{t-i}=e_t+\sum_{j=1}^{l}b_jX_{t-j}e_{t-j}, \tag{4.1.34}$$

and the *lower triangular model*,

$$X_t+\sum_{i=1}^{l}a_iX_{t-i}=e_t+\sum_{\substack{i=1\\ i\geq j}}^{l}\sum_{j=1}^{l}b_{ij}X_{t-i}e_{t-j}. \tag{4.1.35}$$

The diagonal model obviously corresponds to the case where the $\{b_{ij}\}$ matrix in (4.1.1) is purely diagonal, and the lower triangular model similarly corresponds to the case where this matrix is lower triangular. Subba Rao and Gabr (1984, p. 163) have studied the properties of these models in considerable detail, and, in particular, have derived sufficient conditions for their asymptotic stationarity. The treatment of the diagonal model involves rather heavy algebra, and the conditions for stationarity are quite complicated.

Estimation of parameters in bilinear models

Consider a bilinear BL$(p, 0, m, k)$ model of the form

$$X_t + a_1 X_{t-1} + \cdots + a_p X_{t-p} = \alpha + \sum_{i=j}^{m} \sum_{j=1}^{k} b_{ij} X_{t-i} e_{t-j} + e_t \qquad (4.1.36)$$

(This is a particular form of the general bilinear model (4.1.1) in which the pure MA terms have been dropped, and a constant α has been added to the RHS to facilitate the fitting of such models to non mean-corrected data.) The basic principles involved in fitting a model of the form (4.1.36) to data are essentially the same as those involved in conventional ARMA model fitting, and consist of two distinct stages, namely

(1) determining the "orders" of the model, i.e. finding appropriate values for the integers p, m, k; and
(2) for given p, m, k, estimating the paramaters α, $\{a_i\}$, $\{b_{ij}\}$, σ_e^2.

It would appear logical to carry out stage (1) first, followed by stage (2). However, as in linear model fitting, it turns out that the most efficient method of order selection is to fit a range of models involving different values of p, m, k, and then to select the most suitable model on the basis of an "order determination criterion" which takes into account both the magnitude of the estimated residual variance, $\hat{\sigma}_e^2$, and the number of fitted parameters. One well-known criterion is Akaike's AIC (Akaike, 1977), and we shall use this as the basis of the order selection procedure.

Let us now consider stage (2). For a given set of values of p, m, k, and on the assumption that the $\{e_t\}$ are independent zero mean $N(0, \sigma_e^2)$ variables, we can construct the likelihood function of the unknown parameters, given N observations $X_1, X_2, \ldots, X_N$. As with all models which involve lagged values of the $\{X_t\}$, we cannot evaluate the residuals for an initial stretch of the data (see, for example, *Spec. Anal.*, p. 347). We therefore consider the conditional likelihood based on $X_{\gamma+1}, X_{\gamma+2}, \ldots, X_N$, given $X_1, X_2, \ldots, X_\gamma$, where here $\gamma = \max(p, m, k)$.

Now let $\boldsymbol{\theta} = (\theta_1, \theta_2, \ldots, \theta_n)$ denote the complete set of parameters, $\{a_i\}$, $\{b_{ij}\}$, α, i.e. we set

$$\theta_i = a_i, \qquad i = 1, \ldots, p,$$

$$\theta_{p+1} = b_{11}, \qquad \theta_{p+2} = b_{12}, \ldots, \theta_{p+mk} = b_{mk}, \qquad \theta_{p+mk+1} = \alpha,$$

and we write $n = p + mk + 1$ to denote the total number of parameters. The joint probability density function *of* $e_\gamma, e_{\gamma+1}, \ldots, e_N$ is given by

$$(1/2\pi\sigma_e^2)^{(N-\gamma)/2} \exp\left\{ -\frac{1}{2\sigma_e^2} \sum_{t=\gamma+1}^{N} e_t^2 \right\}, \tag{4.1.37}$$

and since the Jacobean of the transformation from $\{X_t\}$ to $\{e_t\}$ is unity, (4.1.37) also represents the likelihood function of $\boldsymbol{\theta}$, given $\{X_t; t = \gamma+1, \ldots, N\}$. The (conditional) maximum likelihood estimates of $\theta_1, \ldots, \theta_n$ are thus given by maximizing (4.1.37), or equivalently by minimizing

$$Q(\boldsymbol{\theta}) = \sum_{t=\gamma+1}^{N} e_t^2. \tag{4.1.38}$$

This minimization has to be performed numerically: for a given set of values of $(\theta_1, \ldots, \theta_n)$ we can evaluate the $\{e_t\}$ recursively from (4.1.36) and then use a numerical "hill-climbing" technique, such as the standard Newton–Raphson method, to minimize $Q(\boldsymbol{\theta})$. The application of this approach to bilinear model fitting has been developed in detail by Subba Rao (1981). The Newton–Raphson iterative equations for the minimizations of $Q(\boldsymbol{\theta})$ are given by,

$$\boldsymbol{\theta}^{(i+1)} = \boldsymbol{\theta}^{(i)} - \boldsymbol{H}^{-1}(\boldsymbol{\theta}^{(i)})\boldsymbol{G}(\boldsymbol{\theta}^{(i)}) \tag{4.1.39}$$

where $\boldsymbol{\theta}^{(i)}$ is the vector of parameter estimates obtained at the ith iteration, and gradient vector $\boldsymbol{G}$ and Hessian matrix $\boldsymbol{H}$ are given by,

$$\boldsymbol{G}(\boldsymbol{\theta}) = \left[\frac{\partial Q}{\partial \theta_1}, \frac{\partial Q}{\partial \theta_2}, \ldots, \frac{\partial Q}{\partial \theta_n} \right]'$$

$$\boldsymbol{H}(\boldsymbol{\theta}) = \left\{ \frac{\partial^2 \theta}{\partial \theta_i \, \partial \theta_j} \right\}.$$

The partial derivatives of Q w.r.t. the $\{\theta_i\}$ are readily shown to be,

$$\frac{\partial Q}{\partial \theta_i} = 2 \sum_{t=\gamma+1}^{N} e_t \frac{\partial e_t}{\partial \theta_i}, \qquad i = 1, 2, \ldots, n,$$

$$\frac{\partial^2 Q}{\partial \theta_i \, \partial \theta_j} = 2 \sum_{t=\gamma+1}^{N} \frac{\partial e_t}{\partial \theta_i} \frac{\partial e_t}{\partial \theta_j} + 2 \sum_{t=\gamma+1}^{N} e_t \frac{\partial^2 e_t}{\partial \theta_i \, \partial \theta_j}, \qquad i, j = 1, \ldots, n.$$

It remains to evaluate the partial derivatives of the $\{e_t)$ w.r.t. the $\{\theta_i\}$, and Subba Rao and Gabr (1984) developed a neat set of recursive equations for those derivatives, as follows. Differentiating (4.1.36) w.r.t. each of the parameters we obtain

$$\frac{\partial e_t}{\partial a_i}+\phi(a_i)=X_{t-i}, \qquad i=1,2,\ldots,p,$$

$$\frac{\partial e_t}{\partial b_{ij}}+\phi(b_{ij})=-X_{t-i}e_{t-j}, \qquad i=1,\ldots,m, \qquad j=1,\ldots,k,$$

$$\frac{\partial e_t}{\partial \alpha}+\phi(\alpha)=-1,$$

where

$$\phi(\theta_l)=\sum_{i=1}^{m}\sum_{j=1}^{k} b_{ij}X_{t-i}\frac{\partial e_{t-j}}{\partial \theta_l}.$$

If we now assume the initial conditions

$$e_t=\frac{\partial e_t}{\partial \theta_i}=0, \qquad t=1,2,\ldots,\gamma, \qquad i=1,\ldots,n,$$

the second-order derivatives satisfy,

$$\frac{\partial^2 e_t}{\partial a_i \partial a_{i'}}=0, \qquad \frac{\partial^2 e_t}{\partial a_i\, \partial \alpha}=0, \qquad i,i'=1,\ldots,p,$$

$$\frac{\partial^2 e_t}{\partial a_i\, \partial b_{jl}}+\psi(a_i,b_{jl})=-X_{t-j}\frac{\partial e_{t-l}}{\partial a_i},$$

$$\frac{\partial^2 e_t}{\partial b_{ij}\, \partial b_{i'j'}}+\psi(b_{ij},b_{i'j'})=-X_{t-i}\frac{\partial e_{t-j}}{\partial b_{i'j'}}-X_{t-i'}\frac{\partial e_{t-j'}}{\partial b_{ij}}$$

$$\frac{\partial^2 e_t}{\partial b_{ij}\, \partial \alpha}+\psi(b_{ij},\alpha)=-X_{t-i}\frac{\partial e_{t-j}}{\partial \alpha},$$

$$\frac{\partial^2 e_t}{\partial \alpha^2}=0,$$

where

$$\psi(\theta_r,\theta_l)=\sum_{i=1}^{m}\sum_{j=1}^{k} b_{ij}X_{t-i}\frac{\partial^2 e_{t-j}}{\partial \theta_r\, \partial \theta_l}.$$

For a given set of parameter values α, $\{a_i\}$, $\{b_{ij}\}$, the first- and second-order derivatives of Q can be evaluated from the above equations, and hence the vector $\boldsymbol{G}$ and matrix $\boldsymbol{H}$ evaluated. The iteration (4.1.39) can now be implemented. When the final parameter estimates, $\hat{\boldsymbol{\theta}}$, have been obtained, σ_e^2 may then be estimated in the usual way by,

$$\hat{\sigma}_e^2 = \frac{1}{N-\gamma} Q(\hat{\boldsymbol{\theta}}) = \frac{1}{N-\gamma} \sum_{t=\gamma+1}^{N} \hat{e}_t^2. \tag{4.1.40}$$

Order selection

As previously remarked, we may determine suitable values of p, m, k by fitting a range of models covering various values of p, m, k, and then selecting that model for which Akaike's criterion, namely

$$\text{AIC} = (N-\gamma) \log \hat{\sigma}_e^2 + 2 \times \text{number of fitted parameters} \tag{4.1.41}$$

attains its minimum value. (See Akaike, 1977); see also *Spec. Anal.*, p. 372 for a general discussion of order determination criteria.) Note that here $(N-\gamma)$ is the effective number of observations to which each model has been fitted, and that in using the AIC criterion we are trying to strike a balance between reducing the magnitude of the residual variance and increasing the number of model parameters.

To implement this approach all we need do, in principle, is set upper bounds to p, m, k, and then search over all combinations of p, m, k within these bounds. However, this would involve searching over a three-dimensional grid, and unless we confine the search to very low values of p, m, k, this procedure would be impracticable. What is clearly required is some form of "nested" search procedure, and Subba Rao (1981) has suggested the following algorithm.

(1) For a given value of p, fit a (linear) AR(p) model.
(2) Using the AR coefficients as initial values for the $\{a_i\}$ and α, and setting initially $b_{11}=0$, fit a BL(p, 0, 1, 1) model using the Newton–Raphson technique described above.
(3) Fit BL(p, 0, 1, 2) and BL(p, 0, 2, 1) models using the parameters of the BL(p, 0, 1, 1) model as initial values and setting initially the remaining bilinear coefficients to zero.
(4) Of the two models fitted in stage (3), choose that model which has the smaller residual variance and use its parameters as starting values of fitting a BL(p, 0, 2,2) model.
(5) The procedure is continued until m, k have reached a common upper bound Γ. At each stage, a bilinear term of order (m, k) is fitted by considering bilinear terms of orders $(m-1, k)$, $(m, k-1)$, and choosing whichever model has the smaller residual variance to provide the

starting values, with initial values for the remaining coefficients set to zero.

(6) All the previous steps are repeated for $p = 1, 2, \ldots, \Gamma$, and the procedure terminates when the residual variance $\hat{\sigma}_e^2$ starts to increase as m, k, increase. (As a working rule, the value of Γ should be at least as large as the order of the best AR model selected by the AIC criterion.) The final choice of model is then made by calculating the AIC values for each fitted model, and selecting that model for which AIC obtains it smallest value.

Subba Rao and Gabr (1984) have considered an alternative method of obtaining initial values for parameter estimates based on a modified version of the *repeated residuals* technique first proposed by Subba Rao (1977) (see also *Spec. Anal.*, p. 881). This method starts also by fitting an AR(p) model to the data, yielding a set of residuals $\{\hat{e}_t^*\}$, say, $t = \gamma+1, \ldots, N$. For given m, k, the parameters of a BL(p, 0, m, k) model are then estimated by minimizing

$$Q(\theta) = \sum_{t=\gamma+1}^{N} \left\{ X_t + \sum_{i=1}^{p} a_i X_{t-i} - \alpha - \sum_{i=1}^{m} \sum_{j=1}^{k} b_{ij} X_{t-i} \hat{e}_{t-j}^* \right\}^2$$

using standard least-squares techniques. The $\{\hat{e}_t^*\}$ are then replaced by the residuals from the fitted BL(p, 0, m, k) model, and the complete procedure is repeated until the parameter estimates converge to a stable set of values. If the estimates fail to converge after a reasonable number of steps, the values of $\hat{\alpha}$, $\{\hat{a}_i\}$, $\{\hat{b}_{ij}\}$ obtained may be used as initial values for the application of the Newton–Raphson method.

Analysis of the sunspot series

As mentioned in Section 3.4.1, the Wolf sunspot in a very well-known data set, and records the annual sunspot index from 1700 onwards. (See *Spec. Anal.*, p. 882) and Morris (1977) for a review of previous studies: the data are given in Appendix 1.) The data are shown in Fig. 4.4 and it will be seen that the main feature of this series is a cycle of activity varying in duration between 9 and 14 years, with an average period of approximately 11.3 years. The series exhibits another feature, namely different gradients of "ascensions" and "descensions", i.e. in each cycle the rise to the maximum has a steeper gradient than the fall to the next minimum. This suggests that a non-linear model might be appropriate, and Haggan and Subba Rao fitted various linear and bilinear models and compared their relative fit. The results of their analyses are given below. In this study the data consisted of 246 observations, with mean 43.53, and all the linear models were fitted

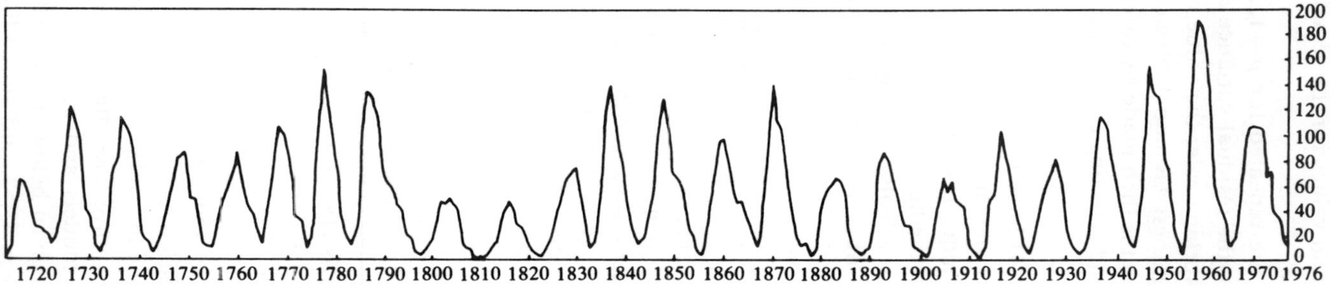

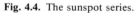

Fig. 4.4. The sunspot series.

to mean-correlated data. The orders of the linear models were selected according to the AIC criterion.

AR models. The best AR model is AR(9) with coefficients,

$$\hat{a}_1 = -1.21, \quad \hat{a}_2 = 0.47, \quad \hat{a}_3 = 0.12, \quad \hat{a}_4 = -0.14, \quad \hat{a}_5 = 0.16,$$
$$\hat{a}_6 = -0.09, \quad \hat{a}_7 = 0.08, \quad \hat{a}_8 = -0.09, \quad \hat{a}_9 = -0.10.$$
$$\hat{\sigma}_e^2 = 194.43, \qquad \text{AIC} = 1316.44.$$

Subset AR models. The best subset AR model includes lags 1, 2, 9, with coefficients

$$\hat{a}_1 = -1.24, \quad \hat{a}_2 = 0.54, \quad \hat{a}_9 = -0.15$$
$$\hat{\sigma}_e^2 = 199.2 \qquad \text{AIC} = 1310.39.$$

ARMA models. The best ARMA model is ARMA(6, 6) with AR coefficients

$$\hat{a}_1 = -0.52, \quad \hat{a}_2 = -0.44, \quad \hat{a}_3 = -0.51, \quad \hat{a}_4 = 1.09,$$
$$\hat{a}_5 = 0.07, \quad \hat{a}_6 = -0.25$$

and MA coefficients,

$$\hat{b}_1 = 0.71, \quad \hat{b}_2 = -0.07, \quad \hat{b}_3 = -1.09, \quad \hat{b}_4 = -0.08,$$
$$\hat{b}_5 = 0.041, \quad \hat{b}_6 = 0.42$$
$$\hat{\sigma}_e^2 = 185.27, \qquad \text{AIC} = 1309.8.$$

Bilinear models. Using the AIC criterion, a BL(p, 0, m, k) model of the form (4.1.36) was chosen with $p = 3$, $m = 3$, $k = 4$. The coefficients are:

$$\hat{\alpha} = 10.9132, \quad \hat{a}_1 = -1.93, \quad \hat{a}_2 = 1.46, \quad \hat{a}_3 = -0.27,$$

and the $\hat{b}_{ij}$ values ($i = 1, 2, 3, j = 1, 2, 3, 4$) are:

-0.55×10^{-2}	0.32×10^{-2}	-0.18×10^{-2}	0.08×10^{-2}
-0.57×10^{-2}	-0.56×10^{-2}	-0.82×10^{-2}	0.58×10^{-2}
-0.17×10^{-2}	0.71×10^{-2}	0.11×10^{-1}	-0.789×10^{-3}

$$\hat{\sigma}_e^2 = 143.86, \qquad \text{AIC} = 1214.58.$$

It will be seen that the bilinear coefficients, $\hat{b}_{ij}$, are small, but their inclusion has substantially reduced both the residual variance and the AIC value. The reduction in residual variance is massively significant when referred to an asymptotic likelihood ratio test (see *Spec. Anal.*, p. 884). Subba Rao and Gabr (1984) have fitted bilinear models of higher order to the first 221 observations, and obtained a BL(8, 0, 5, 4) on the basis of the AIC criterion. This model has residual variance $\hat{\sigma}_e^2 = 110.62$ and AIC value 1051.00.

Finally, we would mention that Subba Rao and Gabr have also fitted the following *subset bilinear model*,

$$\begin{aligned} X_t + \hat{a}_1 X_{t-1} + \hat{a}_2 X_{t-2} + \hat{a}_9 X_{t-9} \\ = \hat{\alpha} + \hat{b}_{2,1} X_{t-2} e_{t-1} + \hat{b}_{8,1} X_{t-8} e_{t-1} + \hat{b}_{1,3} X_{t-1} e_{t-3} \\ + \hat{b}_{4,3} X_{t-4} e_{t-3} + \hat{b}_{1,6} X_{t-1} e_{t-6} + \hat{b}_{2,4} X_{t-2} e_{t-4} + \hat{b}_{3,2} X_{t-3} e_{t-2} + e_t \end{aligned}$$

with

$$\hat{a}_1 = -1.5012, \qquad \hat{a}_2 = 0.7670, \qquad \hat{a}_9 = -0.1152,$$

$$\hat{\alpha} = 6.886, \qquad \hat{b}_{2,1} = -0.1458, \qquad \hat{b}_{8,1} = 0.006312,$$

$$\hat{b}_{1,3} = -0.007152, \qquad \hat{b}_{4,3} = -0.006047, \qquad \hat{b}_{1,6} = 0.003619,$$

$$\hat{b}_{2,4} = 0.004334, \qquad \hat{b}_{3,2} = 0.001782.$$

The residual variance for this subset bilinear model is $\hat{\sigma}_e^2 = 124.33$, which represents a substantial reduction on the residual variance of the full bilinear model.

Subba Rao and Gabr (1984) provide further practical examples of bilinear model fitting, and give a discussion of the sampling properties of the estimated parameters, including some simulation studies of the skewness and kurtosis of their sampling distributions.

Forecasting from bilinear models

Given observations on a series up to time t, it is well known that the minimum mean-square error predictor of a future value X_{t+h} is given by the conditional expectation

$$\tilde{X}(t+h) = E[X_{t+h} \mid X_s;\, s \leq t] \tag{4.1.42}$$

(see Section 7.1). For linear models (with independent residuals) these conditional expectations can be evaluated via a simple set of recursive equations originally proposed by Box and Jenkins (1970)—see also *Spec. Anal.*, p. 762. For bilinear models the situation is somewhat more complicated, but a similar approach can be followed through. Consider the BL(p, 0, m, k) model (4.1.36). Assuming that the model is invertible (so that knowledge of $\{X_s;\, s \leq t\}$ is equivalent to knowledge of $\{e_s;\, s \leq t\}$, and vice versa), and using the fact that X_{t+r} and e_{t+s} are independent for $s > r$, we have immediately,

$$\tilde{X}(t+1) = -a_1 X_t - a_2 X_{t-1} \cdots - a_p X_{t-p+1} + \alpha + \sum_{i=1}^{m} \sum_{j=1}^{k} b_{ij} X_{t+1-i} e_{t+1-j}.$$

For $h > 1$, computation of the conditional expectations is more cumbersome since it involves the evaluation of expressions such as $E\{X_{t+h-i}e_{t+h-j} | X_s; s \le t\}$. However, Subba Rao and Gabr (1984, p. 199) have constructed a recursive algorithm for the evaluation of these expressions. (The difficult case to deal with is $i < j < h$; the essence of Subba Rao and Gabr's technique for this case is to use the model (4.1.36) to express X_{t+h-i} in terms of $\{X_{t+h-k}; k < i\}$.) Using this algorithm we can then compute exact mean-square error predictors for bilinear models.

4.2 THRESHOLD AUTOREGRESSIVE MODELS

A different class of non-linear models, call *threshold models*, was introduced by Tong in a long sequence of papers culminating in Tong and Lim (1980). A detailed account of the theory and application of threshold models is given in the monograph by Tong (1983). The basic idea may be explained as follows. We start with a linear model for a series $\{X_t\}$ and then allow the parameters of the model to vary according to the values of a finite number of past values of X_t, or a finite number of past values of an associated series $\{Y_t\}$. For example, a first-order "threshold autoregressive model" (TAR(1)) would typically be of the form

$$X_t = \begin{cases} a^{(1)}X_{t-1} + e_t^{(1)}, & \text{if } X_{t-1} < d, \\ a^{(2)}X_{t-1} + e_t^{(2)}, & \text{if } X_{t-1} \ge d, \end{cases} \tag{4.2.1}$$

where $\{e_t^{(1)}\}$, $\{e_t^{(2)}\}$ are each strict white noise processes, $a^{(1)}$, $a^{(2)}$, are constants, and the constant d is called the "threshold". This model, which was first introduced by Tong for the analysis of river flow data, is readily extended to an "l-threshold" from, namely

$$X_t = a^{(i)}X_{t-1} + e_t^{(i)} \quad \text{if } X_{t-1} \in R^{(i)}, \qquad i = 1, \ldots, l \tag{4.2.2}$$

where $R^{(1)}, \ldots, R^{(l)}$ are given subsets of the real line R^1 which define a partition of R^1 into disjoint intervals $(-\infty, r_1]$, $(r_1, r_2], \ldots, (r_{l-1}, \infty]$, with $R^{(1)}$ denoting the interval $(-\infty, r]$ and $R^{(l)}$ the interval $(r_{l-1}, \infty]$. The first-order l-threshold model may thus be regarded as a "piecewise linear" approximation to the general non-linear first order model

$$X_t = \lambda(X_{t-1}) + e_t \tag{4.2.3}$$

where $\lambda(\cdot)$ is some general non-linear function. (An extensive discussion of general non-linear first-order models is given by Jones (1978).) Higher-order threshold models may be constructed in a similar way; for example, a TAR(k) model is given by,

$$X_t = a_0^{(i)} + a_1^{(i)}X_{t-1} + \cdots + a_k^{(i)}X_{t-k} + e_t^{(i)} \tag{4.2.4}$$

if $(X_{t-1}, \ldots, X_{t-k}) \in \mathscr{R}^{(i)}$ where $\mathscr{R}^{(i)}$ is a given region of the k-dimensional Euclidean space $\mathscr{R}^k$. The model (4.2.4) may now be regarded as a piecewise linear approximation to the general non-linear kth-order AR model,

$$X_t = f(X_{t-1}, X_{t-2}, \ldots, X_{t-k}) + e_t \tag{4.2.5}$$

Using the state-space representation of AR models (see Section 2.3) the TAR(k) model may be rewritten in the form,

$$\begin{aligned} \boldsymbol{x}_{t+1} &= \boldsymbol{A}^{(i)}\boldsymbol{x}_t + \boldsymbol{B}^{(i)}e_{t+1}^{(i)} \quad \text{if } \boldsymbol{x} \in \mathscr{R}^{(i)}, \\ X_t &= \boldsymbol{H}\boldsymbol{x}_t. \end{aligned} \tag{4.2.6}$$

This may be compared with the state-space representation of the BL(p, 0, p, 1) model (cf. (4.1.3)). The essential distinction is that in the bilinear case the non-linearity is injected by introducing the product term, $e_t\boldsymbol{B}\boldsymbol{x}_t$, whereas in the threshold model the non-linearity arises in terms of the functional relationship between $\boldsymbol{x}_t$ and $\boldsymbol{x}_{t+1}$, with the residual e_{t+1} still entering "linearly". A tenuous link between bilinear and threshold models can be established, as illustrated in the following example. Consider a first-order bilinear system in which the input U_t, output X_t, and noise e_t are related by,

$$X_{t+1} = aX_{t-1} + cU_tX_t + e_{t+1}.$$

If U_t is determined by a feedback mechanism of the form,

$$U_t = \begin{cases} \alpha, & \text{if } X_t < d \\ -\alpha & \text{if } X_t \geq d. \end{cases}$$

then the model for X_t becomes

$$X_{t+1} = \begin{cases} a^{(1)}X_t + e_{t+1}, & \text{if } X_t < d \\ a^{(2)}X_t + e_{t+1}, & \text{if } X_t \geq d, \end{cases}$$

where $a^{(1)} = a + \alpha c$, $a^{(2)} = a - \alpha c$.

In practice it would hardly be feasible to consider fitting to data a model of the form (4.2.4) with k reasonably large since the determination of the threshold regions would involve a search over a k-dimensional space. Tong has therefore restricted attention to the case where the various sets of parameter values are determined by *just a single past value*, X_{t-d}, say. The TAR(k) model then takes the form,

$$X_t = a_0^{(j)} + \sum_{i=1}^{k} a_i^{(j)}X_{t-i} + e_t^{(j)}, \qquad \text{if } X_{t-d} \in R^{(j)}, \qquad j = 1, 2, \ldots, l, \tag{4.2.7}$$

where $R^{(j)}$ is a given subset of the real line R^1.

In (4.2.7) we have assumed a common order, k, for each of the autoregressions; in practice, however, it could well turn out that different threshold regions give rise to models of different orders. (Equation (4.2.7) remains valid since we may take k as the largest order involved, and set all redundant coefficients to zero.) Taking this feature into account we may write (4.2.7) more specifically in the form,

$$X_t = a_0^{(j)} + \sum_{i=1}^{k_j} a_i^{(j)} X_{t-i} + e_t^{(j)}, \qquad \text{if } X_{t-d} \in R^{(j)}, \qquad j = 1, \ldots, l, \quad (4.2.8)$$

where $k_1, k_2, \ldots, k_l$ obviously denote the orders of the autoregressions in the l threshold regions.

Limit cycles

One very interesting feature of threshold models is that, under suitable conditions, they can give rise to *limit cycle* behaviour, thereby mimicking a well-known feature encountered in the study of non-linear differential equations. Thus, if we "switch off" the noise process e_t at time t_0 (i.e. we set $e_t = 0$, $t > t_0$), (4.2.4) may possess a solution $\tilde{X}_t$, which has an *asymptotic periodic form.* Now the solution of the difference equation with $e_t = 0$ is closely related to the "eventual forecast function" of the model. (Note, however, that although $\tilde{X}_{t_0+1} = E[X_{t_0+1} | X_{t_0}, X_{t_0-1}, \ldots]$, in general $\tilde{X}_t \neq E[X_t | X_{t_0}, X_{t_0-1}, \ldots]$, $t > t_0 + 1$.) This means that such models are especially suited to the description of "cyclic" phenomena, and provide an interesting alternative to the more conventional models based on a superposition of harmonic components on a linear stationary residual. In particular, if the series under study exhibits a cyclical form, but with asymmetrical cycles (suggesting the lack of time-irreversibility), then a non-linear model may well provide a more satisfactory description of the data.

Hitherto, the phenomenon of limit cycles has been associated with continuous-time systems. Tong (1983) has proposed the following definition for discrete-time systems.

Let $\boldsymbol{x}_t$ denote a k-dimensional vector $\in R^k$ and satisfying the recurrence relation,

$$\boldsymbol{x}_t = \boldsymbol{f}(\boldsymbol{x}_{t-1}),$$

where $\boldsymbol{f}$ is given a $[k]$ vector-valued function. Let $\boldsymbol{f}^{(j)}$ denote the jth iterate of $\boldsymbol{f}$, i.e. $\boldsymbol{f}^{(j)}(\boldsymbol{x}) = \boldsymbol{f}(\boldsymbol{f}(\ldots \boldsymbol{f}(\boldsymbol{x}) \ldots))$. Then a $[k]$ vector $\boldsymbol{c}_1$ is called a "stable periodic point of period T" with respect to a domain $D \subset R^k$ if for all $\boldsymbol{x}_0 \in D$, $\boldsymbol{f}^{(jT)}(\boldsymbol{x}_0) \to \boldsymbol{c}_1$ as $j \to \infty$, T being the smallest positive integer for which such convergence holds. In this case, $\boldsymbol{c}_1, \boldsymbol{f}^{(1)}(\boldsymbol{c}_1), \boldsymbol{f}^{(2)}(\boldsymbol{c}_1), \ldots, \boldsymbol{f}^{(T-1)}(\boldsymbol{c}_1)$ are all distinct stable limit points of period T, and writing $\boldsymbol{c}_{i+1} = \boldsymbol{f}^{(i)}(\boldsymbol{c}_1)$, $i = 1, 2, \ldots, T-1$, the set of vectors $(\boldsymbol{c}_1, \boldsymbol{c}_2, \ldots \boldsymbol{c}_T)$ is then called a *stable limit cycle of period* T (with respect to D). The case $T = 1$ corresponds to a *stable limit point.*

In the context of time series models, $\boldsymbol{x}_t$ corresponds, of course, to the state-vector. Thus, to demonstrate that a particular model gives rise to a limit cycle we should, ideally, follow through the evolution of the state-vectors as t increases and show that, with an arbitrary starting value, the plot of the state-vectors "spirals in" to form what ultimately becomes a closed curve, namely the "limit cycle". However, for models of the form (4.2.7) this would involve an examination of $[k]$ plots, and with $k>3$ would clearly be impracticable. In empirical investigations of limit cycle behaviour Tong (1983, p. 85) has restricted attention to just the first two components of the state-vector, namely (X_t, X_{t-1}), but in many cases [2] plots of (X_t, X_{t-1}) as t increases are quite informative, and clearly show the form of the limit cycle (when it exists).

Threshold models based on "instrumental processes"

Tong (1983) has extended the class of threshold models to include cases where the "switching" between the different sets of parametric values is determined by a past value of an associated process, $\{Y_t\}$, rather than a past value of $\{X_t\}$. Thus he considers models of the form

$$X_t = a_0^{(j)} + \sum_{i=1}^{m_j} a_i^{(j)} X_{t-i} + \sum_{i=0}^{m_j'} b_i^{(j)} Y_{t-i} + e_t^{(j)}, \tag{4.2.9}$$

if $Y_{t-d} \in R^{(j)}$ $j=1,\ldots,l,$ and (4.2.9) is called a TARSO$\{l, (m_1, m_1'), \ldots, (m_l, m_l')\}$ model (an "open loop threshold autoregressive system").

Bivariate threshold models

If we have a bivariate series (X_t, Y_t) for which X_t satisfies an equation of the form (4.2.9), and also Y_t satisfies

$$Y_t = \alpha_0^{(j)} + \sum_{i=1}^{n_j} \alpha_i^{(j)} X_{t-j} + \sum_{i=1}^{n_j'} \beta_i^{(j)} Y_{t-i} + \eta_t^{(j)} \tag{4.2.10}$$

if $X_{t-d} \in R_j'$, say $(j=1,\ldots,l')$, with $\{\eta_t^{(j)}\}$ similarly a strict white noise process, then (4.2.9), (4.2.10) together constitute a TARSC model (a "closed loop threshold autoregressive system").

Tong has also introduced a "canonical form" for autoregressive threshold models, namely

$$X_t = a_0^{(J_t)} + \sum_{i=1}^{k} a_i^{(J_t)} X_{t-i} + h^{(J_t)} e_t \tag{4.2.11}$$

where $\{J_t\}$ is a general "indicator process" taking integer values, i.e. $J_t = 1, 2, \ldots, l$, and conditional on $J_t = j$, $a_i^{(J_t)} = a_i^{(j)}$, etc. The TAR(k) model (4.2.7) now corresponds to the case where $\{J_t\}$ is a function of $\{X_{t-d}\}$, i.e. a function of a past value of the given series. To emphasize this point, Tong calls (4.2.7) a SETAR ("self-exciting threshold autoregressive model"). In particular, (4.2.8) is referred to as a SETAR $(l, k_1, k_2, \ldots, k_l)$ model.

EAR models

To illustrate a rather different application of (4.2.11), well outside the context of threshold models, we refer to the EAR ("exponential autoregressive") models introduced by Lawrence and Lewis—see, for example Lawrence and Lewis (1985). Their second-order model may be written as

$$X_t = \begin{cases} \beta_1 X_{t-1}, & \text{with probability } \alpha_1 \\ \beta_2 X_{t-2}, & \text{with probability } \alpha_2 \\ 0, & \text{with probability } 1-\alpha_1-\alpha_2. \end{cases} + e_t$$

However, these models can equally well be written in the form

$$X_t = a^{(J_t)} X_{t-1} + b^{(J_t)} X_{t-2} + e_t,$$

where the process $\{J_t\}$ is *independent* of $\{X_t\}$, with

$$J_t = \begin{cases} 0, & \text{with probability } \alpha_1 \\ 1, & \text{with probability } \alpha_2 \\ 2, & \text{with probability } 1-\alpha_1-\alpha_2 \end{cases}$$

and $a^{(0)} = \beta_1$, $b^{(0)} = 0$, $a^{(1)} = 0$, $b^{(1)} = \beta_2$, $a^{(2)} = b^{(2)} = 0$. (See Priestley (1985) for further discussion of this representation of EAR models.)

Structural properties of threshold models

Tong (1983, p. 93) has derived sufficient conditions for the ergodicity of a first-order threshold model, i.e. a SETAR $(l, 1, 1, \ldots, 1)$ of the form

$$X_t = a_0^{(j)} + a_1^{(j)} X_{t-1} + e_t, \qquad \text{if } X_{t-1} \in R^{(j)}, \qquad j = 1, 2, \ldots, l.$$

Tong shows that a process satisfying the above model (with e_t having an absolutely continuous distribution) is ergodic if $|a_1^{(1)}| < 1$ and $|a_1^{(l)}| < 1$, these results following from more general results on the ergodic properties of the general, first-order, non-linear model (4.2.3) due to Jones (1978). (Note that $a_1^{(1)}$, $a_1^{(l)}$ denote respectively the coefficients of the AR(1) models fitted over the extreme left and extreme right intervals of R^1.) Tong has also studied the stationary (marginal) distributions of first-order threshold models, and shows how one may obtain the MGF of the distribution of $\{e_t\}$, given the MGF of the stationary distribution of $\{X_t\}$. However, the dual problem, namely determining the distribution of $\{X_t\}$ from a given distribution for $\{e_t\}$ seems more intractable, and Tong leaves the solution in the form of an integral equation which does not lend itself to an analytic solution.

There appear to be few, if any, results available for the theoretical autocovariance and spectral density functions of general stationary threshold models. (Tong (1983, p. 101) shows how some results of Jones (1978) for general, first-order, non-linear models can be used to derive the autocovariance function of a first-order SETAR model.) When fitting threshold models to data it is often useful to be able to compare the sample autocovariance function of the data with the autocovariance function generated by the fitted model. To effect such comparison, Tong uses the model to generate artificial data, and then "estimates" the autocovariance function of the model from these artificial data.

Estimation of parameters in threshold models

Given N observations $(X_1, X_2, \ldots, X_N)$ from a series $\{X_t\}$, a SETAR model of the form (4.2.8) may be fitted by fitting each of the l component models separately to the appropriate subset of the observations. For example, if we focus attention on the jth component model we select that subset of the observations, $X_{t_1}, X_{t_2}, \ldots, X_{t_M}$, say, for which $X_{t_i-d} \in R^{(j)}$, all i, and then estimate the coefficients $a_0^{(j)}, a_1^{(j)}, \ldots, a_{k_j}^{(j)}$ by using standard least-squares (or equivalently maximum likelihood under the assumption of Gaussianity)—exactly as in the case of linear AR model fitting. The estimation of the coefficients thus presents little difficulty. However, the determination of the "structural parameters", namely the "delay parameter", d, the threshold region, $\{R^{(j)}\}$, and the individual model orders, $k_1, \ldots, k_l$, is a more difficult problem. The algorithm proposed by Tong (1983, p. 131) rests heavily on the use of Akaike's AIC criterion (cf. (4.1.41)), and its operation is illustrated on the following SETAR$(2, k_1, k_2)$ (i.e. single threshold) model,

$$X_t = \begin{cases} a_0^{(1)} + \sum_{i=1}^{k_1} a_1^{(1)} X_{t-i} + e_t^{(1)}, & \text{if } X_{t-d} \geq r, \\ a_0^{(2)} + \sum_{i=1}^{k_2} a_i^{(2)} X_{t-1} + e_t^{(2)}, & \text{if } X_{t-d} < r. \end{cases}$$

The algorithm proceeds as follows.

Stage 1. For given values of d, r, fit separate AR models to the appropriate subsets of the data. Let AIC(k_1), AIC(k_2), denote the usual AIC criteria for the individual models, and let $\hat{k}_1$, $\hat{k}_2$ denote these values which minimize AIC(k_1), AIC(k_2), respectively. Write

$$\text{AIC}(d, r) = \text{AIC}(\hat{k}_1) + \text{AIC}(\hat{k}_2).$$

Stage 2. Consider a set of possible value for r, say, $r^{(1)}, r^{(2)}, \ldots, r^{(q)}$. (Typically these will be percentiles of the range of variation of the $\{X_t\}$.) Repeat stage 1 for $r = r^{(i)}$, $i = 1, \ldots, q$, with d remaining fixed. Choose that value, $\hat{r}$, say, for which AIC(d, r) attains its minimum value, and write

$$\text{AIC}(d) = \text{AIC}(d, \hat{r}).$$

Stage 3. Now search for the "best" value of d over a range of possible values, $d_1, d_2, \ldots, d_p$, say, by repeating *both* stages 1 and 2 for $d = d_i$, $i = 1, \ldots, p$. Select that value of d for which AIC(d) attains its minimum value. (Since changes in the value of d may also change the "effective number of observations" to which the model is fitted, the AIC criterion used in stage 3 is the "normalized version", i.e. the standard AIC value divided by the effective number of observations.)

Tong argues that in determining the final choice of model one need not adhere strictly to the values of the "structural parameters" selected by the AIC criteria. Rather, the AIC criteria are used as a guide to select a relatively small subclass of plausible models which may then be examined for certain special properties (such as limit cycle behaviour) which other considerations suggest are desirable features. Tong (1983) has also compiled an extensive set of "diagnostic" checks aimed at assessing whether a fitted model shares the main characteristics of the data. These checks include: examining the data for cyclical behaviour; plotting univariate and bivariate histograms of both the raw data and artificial data generated from the model; comparing spectral density functions and higher-order spectra estimated from the raw data and from the artificial data; computing "kernel type" estimates of the regressions functions of X_t on X_{t-j} $(j = 1, \ldots, p)$; and testing the residuals from the fitted model to see whether they are plausibly both Gaussian and "white". A further check often used is to fit the model to only part of the given data, and then compare the remaining observations with the corresponding values predicted by the model.

Distribution of coefficient estimates

Using some general results of Klimko and Nelson (1978) on conditional least-squares, Tong (1983) shows that for an ergodic SETAR($2, k_1, k_2$) model the set of estimated coefficients has an asymptotic multivariate normal distribution. (This result does not take account of the sampling properties of the estimates of the *structural parameters*, but assumes that these are known, *a priori*). However, the computation of the variance–covariance matrix of the multivariate normal distribution is extremely complicated (except for some very simple models), and Tong suggests that, as an approximation, one may compute the variance–covariance matrix of the estimated coefficients for the jth component model by applying standard least-squares theory to that subset of the observations appropriate to the jth model. In those cases where the asymptotic variance–covariance matrix can be evaluated, simulation results indicate that the two methods are in close agreement, and the least-squares approach is the one commonly used.

Forecasting from threshold models

As noted in Section 4.1, given observations up to time t, the minimum mean-square error predictor of X_{t+h} is given by the conditional expectation (4.1.42). Suppose now that the observations conform to a SETAR model of the form (4.2.8). If $h \leq d$ (the delay parameter) then X_{t+h-d} belongs to the given observations, and thus we know which of the l component models governs X_{t+h} (and similarly which component models govern X_{t+h-1}, $X_{t+h-2}, \ldots, X_{t+1}$). Under the assumption of independent $\{e_t\}$, the conditional expectation for $\tilde{X}(t+h)$ can then be evaluated via a set of recursive equations which are immediately obtained using exactly the same techniques as that originally proposed by Box and Jenkins (1970) for linear models. However, if $h > d$ the situation becomes more complicated, and the evaluation of the conditional expectations is intractable. One obvious *ad hoc* device is to set $e_{t+1} = e_{t+2} = \cdots = e_{t+h} = 0$, and simply use the "deterministic" part of the model to generate the predictors. (This is essentially equivalent to replacing any unknown $\{X_{t-d}\}$ by their predictors $\{\hat{X}_{t-d}\}$ previously obtained.)

For cyclical data with period T, say, Tong (1983) suggests that a more efficient procedure is to replace the conditioning variable X_{t-d} by X_{t-T-d}, and then refit the threshold model using the new conditioning variable. This then enables us to compute conditional expectations (from the modified model) for $h \leq T + d$. If still longer-term predictors are required, we may repeat this technique and fit a new threshold model based on the conditioning variable X_{t-2T-d}, etc. A number of numerical illustrations of this approach are given in Tong (1983).

Numerical examples

Sunspot series

Tong and Lim (1980) fitted a SETAR(2, 4, 12) model to the period 1700–1920 (the maximum possible order being fixed at 20). The conditioning variable is X_{t-3}, and the threshold level is 36.6. The detailed model is as follows:

$$X_t = \begin{cases} 10.54+1.69X_{t-1}-1.16X_{t-2}+0.24X_{t-3}+0.15X_{t-4}+e_t^{(1)}, \\ \qquad\qquad\qquad\qquad\qquad\qquad\qquad\qquad \text{if } X_{t-3} \leq 36.6, \\ 7.80+0.74X_{t-1}-0.04X_{t-2}-0.20X_{t-3}+0.17X_{t-4} \\ -0.23X_{t-5}+0.02X_{t-6}+0.16X_{t-7}-0.26X_{t-8}+0.32X_{t-9} \\ +0.39X_{t-10}+0.43X_{t-11}-0.04X_{t-12}+e_t^{(2)}, \qquad \text{if } X_{t-3} > 36.6, \end{cases}$$

with $\hat{\sigma}^2_{e^{(1)}} = 254.64$, $\hat{\sigma}^2_{e^{(2)}} = 66.80$, and "pooled" residual variance = 153.7. The residual variance for this model is comparable with that of the full bilinear model ($\hat{\sigma}_e^2 = 142.86$) given in Section 4.1, but somewhat larger than the residual variance of the subset bilinear model ($\hat{\sigma}_e^2 = 124.33$). An interesting feature of the fitted SETAR model is that the "eventual forecasting function" produces an asymmetric limit cycle with period 31 years. The limit cycle consists of three subcycles whose "rise" and "fall" times are (4, 6), (4, 6), and (4, 7) years, respectively. This asymmetry in the limit cycle agrees well with similar features in the raw data. (In an addendum to his 1983 monograph, Tong points out that an improved computer program leads to a SETAR(2, 3, 11) model. The conditioning variable and the threshold level are as given above, and the residual variances are also essentially unchanged.)

Canadian lynx series

As noted in Section 3.4.1, the Canadian lynx series consists of 114 observations describing the annual number of lynx trapped in the Mackenzie river district of Canada over the years 1821–1934. (For a review of previous analyses of this series see *Spec. Anal.*, p. 384; the full data are given in Appendix 1.) The data, logarithmically transformed, are shown in Fig. 4.5. Like the sunspot series, this is another celebrated data set and, in common with the sunspot series, it also exhibits strong cyclical behaviour. Here, the period is approximately 10 years, and again there is a marked asymmetry in the individual cycles with a "rise" time varying between 5 and 7 years and a "fall" time of 3 to 4 years. (Note that the asymmetry in the lynx series cycles is in the "opposite direction" to that of the sunspot series: the sunspot series has a short "rise" time and a long "fall" time.) This asymmetry again suggests the possibility of non-linear modelling. Tong and Lim (1980)

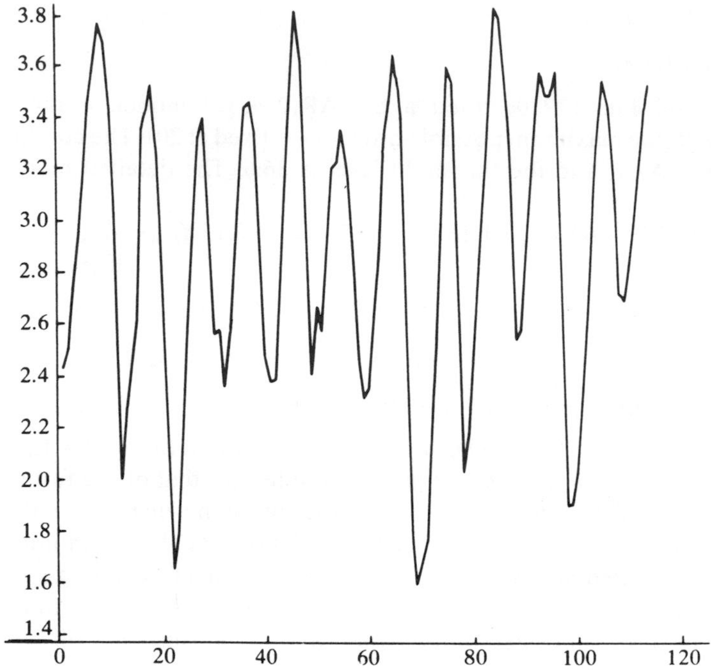

Fig. 4.5. The Canadian lynx series.

fitted the following SETAR(2, 8, 3) model to the logarithmically transformed (to base 10) series covering the full period of 114 observations;

$$X_t = \begin{cases} 0.524 + 1.036X_{t-1} - 0.176X_{t-2} + 0.175X_{t-3} - 0.434X_{t-4} \\ \quad +0.346X_{t-5} - 0.303X_{t-6} + 0.217X_{t-7} + 0.004X_{t-8} + e_t^{(1)}, \\ \qquad\qquad\qquad\qquad\qquad\qquad\qquad\qquad \text{if } X_{t-2} \leq 3.116, \\ 2.656 + 1.425X_{t-1} - 1.162X_{t-2} - 0.109X_{t-3} + e_t^{(2)} \quad \text{if } X_{t-2} > 3.116, \end{cases}$$

with $\hat{\sigma}^2_{e^{(1)}} = 0.0255$, $\hat{\sigma}^2_{e^{(2)}} = 0.0516$, and "pooled" residual variance $= 0.0360$.

The "eventual forecasting function" for this model shows an *asymmetric limit cycle* with period 9 years, a "rise" time of 6 years, and a "fall" time of 3 years. Both the asymmetry and the periodicity are in good agreement with the data. (In the addendum to his monograph, Tong gives a slightly modified version of the above model in the form of a SETAR(2, 7, 2) scheme. The threshold level and residual variances are essentially unchanged.)

By way of comparison, Subba Rao and Gabr (1984) fitted the following subset bilinear model to the first 100 observations (covering the period

1821–1920), again using the logarithmically transformed data:

$$X_t + \hat{a}_1 X_{t-1} + \hat{a}_2 X_{t-2} + \hat{a}_3 X_{t-3} + \hat{a}_4 X_{t-4} + \hat{a}_9 X_{t-9} + \hat{a}_{12} X_{t-12}$$
$$= \hat{\alpha} + \hat{b}_{3,9} X_{t-3} e_{t-9} + \hat{b}_{9,9} X_{t-9} e_{t-9} + \hat{b}_{6,2} X_{t-6} e_{t-2}$$
$$+ \hat{b}_{1,1} X_{t-1} e_{t-1} + \hat{b}_{2,7} X_{t-2} e_{t-7} + \hat{b}_{4,2} X_{t-4} e_{t-2} + e_t$$

with

$$\hat{a}_1 = -0.77277, \quad \hat{a}_2 = 0.09157, \quad \hat{a}_3 = -0.08307$$
$$\hat{a}_4 = 0.26149, \quad \hat{a}_9 = -0.22559, \quad \hat{a}_{12} = -0.24584$$
$$\hat{\alpha} = -1.48629, \quad \hat{b}_{3,9} = -0.7893, \quad \hat{b}_{9,9} = 0.4798,$$
$$\hat{b}_{6,2} = 0.3902, \quad \hat{b}_{1,1} = 0.1326, \quad \hat{b}_{2,7} = 0.07944$$
$$\hat{b}_{4,2} = -0.3212.$$

The residual variance is $\hat{\sigma}_e^2 = 0.0223$. The subset bilinear model, having a lower residual variance, produces a smaller mean-square error for the one-step ahead predictors, but, as observed by Subba Rao and Gabr, the threshold model performs better in terms of prediction error for up to five steps ahead predictors.

It is interesting to compare both the above non-linear models with that fitted by Campbell and Walker (1977). Campbell and Walker considered a "mixed spectrum" model of the form

$$X_t = \mu + A \cos \omega_0 t + B \sin \omega_0 t + Y_t$$

where $\{Y_t\}$ is a (linear) AR(2) process. Their analysis gave $\hat{\omega}_0 = 2\pi/9.63$, corresponding to a period of 9.63 years,

$$\hat{\mu} = 2.9036, \qquad \hat{A} = 0.0895, \qquad \hat{B} = -0.6249$$

and

$$Y_t = 0.971\, Y_{t-1} - 0.2654\, Y_{t-2} + e_t,$$

with $\hat{\sigma}_e^2 = 0.042$.

(Note that for this model the "eventual forecast function" would consist of a pure "sine wave" and would give rise to *symmetrical* cyclical behaviour.) The threshold model thus gives a slightly smaller residual variance, and a more realistic form of asymptotic cyclical structure.

4.3 EXPONENTIAL AUTOREGRESSIVE MODELS

Exponential autoregressive models were introduced by Ozaki (1978) and Haggan and Ozaki (1981) in an attempt to construct time series models which reproduce certain features of non-linear random vibrations theory. An account of recent developments in this area is given by Ozaki (1982, 1985). (Note that although the terminology is similar, these models are quite different from the EAR models discussed by Lawrence and Lewis (1985) and referred to briefly in Section 4.2.)

Physical characteristics of non-linear systems

Non-linear random vibrations are typically described by second-order differential equations of the form

$$\ddot{x}(t)+f\{\dot{x}(t)\}+g\{x(t)\}=y(t)$$

where $f\{\cdot\}$ (the "damping force") and $g\{\cdot\}$ (the "restoring force") are non-linear functions, and $y(t)$ is a stochastic "driving force" input (see, for example, Stoker, 1950). Two examples are:

(i) *Duffing's equation*

$$\ddot{x}(t)+c\dot{x}(t)+ax(t)+b\{x(t)\}^3=y(t) \tag{4.3.1}$$

(ii) *Van der Pol's equation*

$$\ddot{x}(t)+f\{\dot{x}(t)\}+ax(t)=y(t). \tag{4.3.2}$$

If f and g are both *linear* functions, the system equation takes the form

$$\ddot{x}(t)+\alpha_1\dot{x}(t)+\alpha_2 x(t)=y(t) \tag{4.3.3}$$

say. If we now set $y(t)=e(t)$ (white noise), then $x(t)$ becomes a (continuous time) AR(2) process and its (non-normalized) spectral density function is given by (*Spec. Anal.*, p. 278)

$$h_x(\omega)=\frac{C}{|-\omega^2+i\alpha_1\omega+\alpha_2|^2}, \tag{4.3.4}$$

C being a constant. For suitable values of α_1, α_2, $h_x(\omega)$ has a "peak" at a non-zero frequency ω_1, say, and when the system is "driven" by a white noise input the output, $x(t)$, exhibits an approximate periodic form, with period $2\pi/\omega_1$. Note that here the "*resonant frequency*" ω_1 *depends purely on the values of* α_1, α_2, and remains constant over time.

Also, if the input is periodic, i.e. if $y(t)=A\cos\omega_0 t$, then it is well known that the output is also periodic, with the same frequency, but with a modified amplitude; specifically we have

$$x(t)=A|\Gamma(\omega_0)|\cos(\omega_0 t+\phi), \tag{4.3.5}$$

where $|\Gamma(\omega)|=|-\omega_0^2+i\alpha_1\omega_0+\alpha_2|^{-1}$ is a *continuous function* of ω_0. ($\Gamma(\omega)$ is

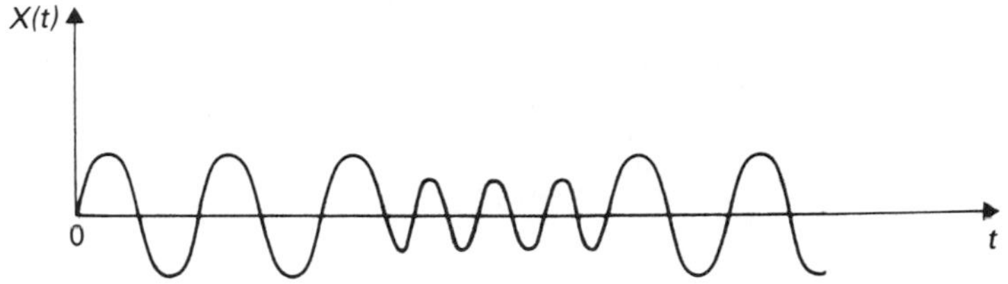

Fig. 4.6. Illustration of amplitude-dependent frequency: case $b<0$.

the "transfer function" of the system—see *Spec. Anal.*, p. 287.) The amplitude of the output thus *changes continuously* as we vary ω_0.

Non-linear systems, however, exhibit certain features which do not occur in the linear case. Three important non-linear features are summarized below.

1. Amplitude-dependent frequency

If a non-linear system is "driven" by white noise the output frequency (ω_1) *may vary with the local amplitude of oscillation*, i.e. ω_1 may take different values depending on whether $x(t)$ is locally "small" or "large". Consider, for example, Duffing's equation (4.3.1) with $y(t)\equiv e(t)$. If $b>0$ (the so-called "hard spring" case) the output frequency increases as the local amplitude increases, whereas if $b<0$ (the "soft spring" case) the frequency decreases as the local amplitude increases (see, for example, Crandall, (1963)). The form of $x(t)$ for the case $b<0$ is indicated in Fig. 4.6.

2. Jump phenomena

Consider a non-linear system governed by Duffing's equation (4.3.1) and driven by a periodic input with frequency ω_0, i.e. we have,

$$\ddot{x}(t)+c\dot{x}(t)+ax(t)+b\{x(t)\}^3=A\cos\omega_0 t \tag{4.3.6}$$

It can be shown that an approximate solution of (4.3.6) is given by (Cunningham, 1958; Stoker, 1950):

$$x(t)\doteqdot B\cos(\omega_0 t+\phi),$$

i.e. the output is approximately periodic with the same frequency ω_0, but now *the amplitude B is a discontinuous function of* ω_0. As ω_0 passes through certain frequencies, B "jumps" from one value to another. This feature is illustrated in Figs 4.7, 4.8 (for $b>0$ and $b<0$, respectively), and it will be seen from Fig. 4.7 that as ω_0 increases to ω_1, the value of B "jumps" to the lower branch of the curve. Similarly, as ω_0 decreases to ω_2, B "jumps" to

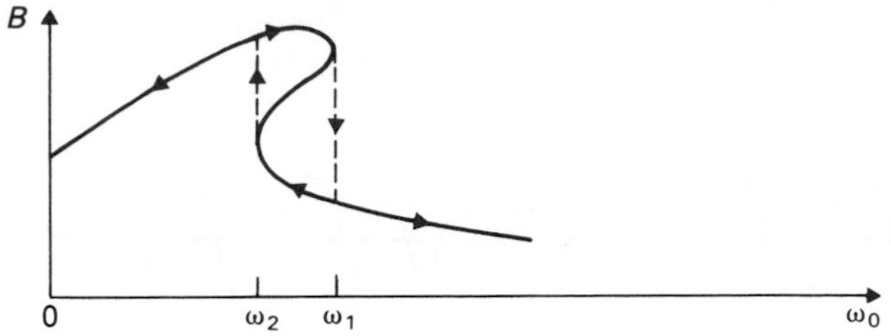

Fig. 4.7. Illustration of "jump" phenomenon: case $b > 0$.

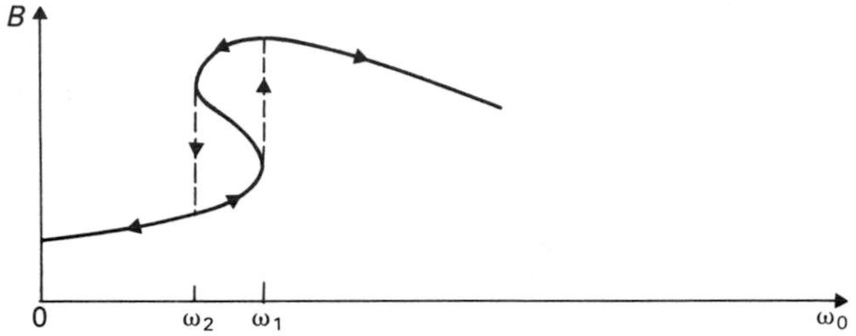

Fig. 4.8. Illustration of "jump" phenomenon: case $b < 0$.

the upper branch of the curve. Note that, in general, ω_1 and ω_2 are different frequencies—corresponding to a type of "hysteresis" effect. For $b > 0$ (Fig. 4.8), the reverse situation holds. (This phenomenon, whereby continuous variation in a "control" variable results in discontinuous variation in another variable is a well-known characteristic in "*catastrophe theory*".)

3. *Limit cycles*

In Section 4.2 we described the notion of "limit cycles" for discrete time systems, and it was pointed out that this type of behaviour is well known in the study of continuous time systems governed by non-linear differential equations. Consider Van der Pol's equation, (4.3.2), with zero input, i.e.

$$\ddot{x}(t) + f\{\dot{x}(t)\} + ax(t) = 0. \tag{4.3.7}$$

The solution of (4.3.7) may form a limit cycle, i.e. execute "self-excited oscillations" which asymptotically become periodic. Specifically, if

$$\dot{x}f(\dot{x}) < 0, \qquad \text{for small } |\dot{x}|,$$

and

$$\dot{x}f(\dot{x}) > 0, \qquad \text{for large } |\dot{x}|,$$

then a limit cycle occurs. In this case the "damping" is negative for small $|\dot{x}|$, i.e. the system "absorbs" energy and the amplitude increases, but for large $|\dot{x}|$ the system "dissipates" energy and the amplitude decreases. (These features arise in the study of acoustics (Rayleigh, 1945), and electrical circuit theory (Van der Pol, 1927).)

Exponential AR models

In order to construct time series models (in discrete time) which reproduce the effect of "amplitude-dependent frequency", Ozaki (1978) started by taking, for example, an AR(2) model of the form

$$X_t - a_1 X_{t-1} - a_2 X_{t-2} = e_t \tag{4.3.8}$$

and then allowed the coefficients a_1, a_2, to depend on X_{t-1}. Specifically, he proposed that the coefficients be made *exponential functions* of X_{t-1}^2, i.e. take the form

$$a_1 = \phi_1 + \pi_1 \exp(-\gamma X_{t-1}^2) \tag{4.3.9}$$

$$a_2 = \phi_2 + \pi_2 \exp(-\gamma X_{t-1}^2). \tag{4.3.10}$$

With these values of a_1 and a_2, (4.3.8) is called a (second-order) *exponential autoregressive model.*

If we ignore, for the moment, the fact that a_1 and a_2 are functions of X_{t-1}, and think of (4.3.8) as a "linear" model, then its "resonant frequency" will occur at the minimum of

$$|1 - a_1 e^{-i\omega} - a_2 e^{-2i\omega}|,$$

and hence *will change with the magnitude of* $|X_{t-1}|$. (In effect, we are assuming that "locally" X_t behaves as if it were generated by a linear AR(2) model with the coefficients a_1, a_2, "frozen" at the values which they attained at time *t*.)

Note that, for large $|X_{t-1}|$,

$$a_1 \sim \phi_1, \qquad a_2 \sim \phi_2,$$

and for small $|X_{t-1}|$,

$$a_1 \sim \phi_1 + \pi_1, \qquad a_2 \sim \phi_2 + \pi_2.$$

Thus, the exponential AR model behaves rather like the threshold AR model, but here the coefficients change "smoothly" between the two extreme values.

In addition to generating "amplitude-dependent frequency" effects, the exponential AR model can also give rise to "jump phenomena" and limit cycle behaviour (Ozaki, 1978). It should be noted, however, that the class of exponential AR models is not unique in these respects: threshold autoregressive models are also capable of generating amplitude-dependent frequency, "jump" phenomena, and limit cycles (Tong, 1983, p. 77).

General model

The second-order model (4.3.8) may be readily extended to a general-order model. Thus, a kth-order exponential AR model is given by,

$$X_t = (\phi_1 + \pi_1\, e^{-\gamma X_{t-1}^2})X_{t-1} + \cdots + (\phi_k + \pi_k\, e^{-\gamma X_{t-1}^2})X_{t-k} + e_t. \quad (4.3.11)$$

Ozaki (1978) has shown that necessary conditions for the existence of a limit cycle for (4.3.11) are:

(1) all the roots of

$$z^k - \phi_1 z^{k-1} \ldots - \phi_k = 0$$

lie inside the unit circle, $|z| = 1$; and

(2) some of the roots of

$$z^k - (\phi_1 + \pi_1)z^{k-1} \cdots - (\phi_k + \pi_k) = 0$$

lie outside the unit circle.

A sufficient condition for the existence of a limit cycle is then

(3) $$\frac{1 - \sum_i \phi_i}{\sum_i \pi_i} > 1 \text{ or } < 0.$$

(The last condition is required to prevent the occurrence of a stable singular point.)

Estimation of parameters is exponential AR models

Haggan and Ozaki (1981) give the following procedure for estimating the parameters of the general model (4.3.11):

1. First fix the value of γ: then $\phi_1, \pi_1, \phi_2, \pi_2, \ldots, \phi_k, \pi_k$, may be estimated by standard least-squares regression analysis of X_t on X_{t-1}, $X_{t-1} \exp(-\gamma X_{t-1}^2), \ldots$. The order k is determined by minimizing Akaike's AIC criterion (see Section 4.1).
2. The above analysis is repeated using a range of values of γ, and the AIC criterion is used to select the most suitable value of γ. The values of γ selected are such that $\exp(-\gamma X_{t-1}^2)$ varies reasonably widely over the range $(0, 1)$.

Numerical example

Haggan and Ozaki (1981) fitted an exponential AR model of the form (4.3.11) to the logarithmically transformed (and mean-connected) Canadian lynx data over the years 1821–1934. Using the procedure described above, they found that the best order was $k=11$, with $\hat{\gamma}=3.89$. The estimated values of $\{\phi_i\}$, $\{\eta_i\}$ were:

i:	1	2	3	4	5	6	7	8	9	10	11
$\hat{\phi}_i$	1.09	−0.28	0.27	−0.45	0.41	−0.36	0.22	−0.10	0.22	−0.07	−0.38
$\hat{\pi}_i$	0.01	−0.49	−0.06	0.30	−0.54	0.61	−0.53	0.30	−0.18	0.18	0.16

The residual variance for this model is $\hat{\sigma}_e^2=0.0321$, which is very close to that given by Lim and Tong's SETAR(2, 8, 3) model, for which the pooled residual variance is 0.0360. The exponential AR model also compares favourably with the "mixed spectrum" model of Campbell and Walker (1977) for which $\hat{\sigma}_e^2=0.042$.

For the above exponential AR model we have:

(1) the roots $z^{11}-\phi_1 z^{10}\cdots-\hat{\phi}_{11}=0$ all lie inside the unit circle; and
(2) two roots of $z^{11}-(\hat{\phi}_1+\hat{\pi}_1)z^{10}\cdots-(\hat{\phi}_{11}+\hat{\pi}_{11})=0$ lie outside the unit circle.

Also,

$$\left\{1-\sum_i \hat{\phi}_i\right\}\Big/\sum_i \hat{\pi}_i=-1.16,$$

and hence a limit cycle occurs. In fact, the above model exhibits a limit cycle with an approximate period of 10 years: for small $|X_{t-1}|$ the period is approximately 10.1 years, while for large $|X_{t-1}|$ the period is approximately 9.4 years. These features agree well with the corresponding effects observed in the raw data.

Chapter 5

General State-dependent Models

The fitting of non-linear models to time series data raises some quite complex issues. In the preceding chapter we discussed three particular types of non-linear models, namely the bilinear, threshold autoregressive, and exponential autoregressive models. Each of these special non-linear schemes describes, in its own way, a specific type of non-linearity, but in practice it may be extremely difficult to decide which, if any, of these special models is best suited to a particular set of data. (In some respects, the situation is similar to that encountered in the study of non-linear differential equations: there is only one type of linear differential equation, but an equation can be non-linear in an infinite variety of ways.) These considerations raise the question as to whether it might be possible to construct a *general class* of non-linear models, which includes the special models of Chapter 4 as special cases, but also allows for a more flexible form of non-linear structure, and, more importantly, *is amenable to statistical analysis.* Now we have already seen that the general (non-anticipative) non-linear model, (3.1.1), is quite intractable as far as statistical analysis is concerned. We are thus faced with the problem of striking a judicious balance between generality and tractability. To highlight this point we may note that (3.1.1) is completely general but totally intractable, whereas, for example, the bilinear model (4.1.4) is certainly tractable but constrains us to a particular form of non-linear structure.

Our approach in this chapter is to start from the general model (3.1.1), and then impose further restrictions until we have reached the stage at which statistical analysis may commence. We must not, of course, impose restrictions which specify a particular parametric form for the function h in (3.1.1)—this would defeat the main object of the exercise. Rather, we will argue in terms of "smoothness" and "analytic" properties, and introduce the crucial assumption of "finite dimensionality" of the model.

These ideas lead to a general class of non-linear models called *state-dependent models* (*SDM*), which, as indicated above, includes the bilinear, threshold autoregressive, and exponential autoregressive models as special cases, but allows much greater flexibility in the character of the non-linear structure. Although these state-dependent models are of a general nature, we will see that they are nevertheless amenable to statistical analysis, and may be fitted to data. This approach to non-linear time series analysis offers two major advantages, namely (i) state-dependent models may be used directly in connection with, for example, the problem of forecasting, and (ii) since they can be fitted to data without any specific prior assumptions about the form of the non-linearity, they may be used to give us an "overview" of the character of the non-linearity inherent in the data, and thus indicate whether, for example, a bilinear, threshold AR, exponential AR model, or even a linear model, is appropriate.

State-dependent models were introduced by Priestley (1980), and their application to various data sets is described in Haggan *et al.* (1984), Priestley and Heravi (1985), Cartwright and Newbold (1983), and Cartwright (1984, 1985).

5.1 CONSTRUCTION OF STATE-DEPENDENT MODELS

Let us return to the general non-linear model (3.1.3), together with its "dual" form (3.1.2). As it stands, (3.1.1) is "infinite dimensional" in the sense that it involves a relationship between infinitely many variables. We now reduce this to a "finite dimensional" form by assuming that the relationship between X_t and the "past history" of the series can be described in terms of finitely many values of past $\{X_t\}$ and $\{e_t\}$, so that we may write, say,

$$X_t = h(X_{t-1}, \ldots, X_{t-k}, e_{t-1}, \ldots, e_{t-l}) + e_t. \qquad (5.1.1)$$

(Note that the function h in (4.1.1) is not, of course, the same function as that which appears in (3.1.1), but it is convenient to retain the same notation.)

With this formulation e_t plays the role of the *innovations* process for X_t, and the function h describes the information on X_t contained within its past history. More formally, we may interpret h as a *conditional expectation*, as follows. Let $\mathscr{B}_t^{(e)} = \mathscr{B}_t(e_t, e_{t-1}, \ldots)$ be the sequence of Borel fields generated by $e_t, e_{t-1}, \ldots$. Then we may write

$$h(X_{t-1}, \ldots, X_{t-k}, e_{t-1}, \ldots, e_{t-l}) = E[X_t \mid \mathscr{B}_{t-1}^{(e)}].$$

Thus, in postulating the model (5.1.1) we are, in effect, saying that the "projection" of X_t on $\mathscr{B}_{t-1}^{(e)}$ *(in the sense of conditional expectation) can be expressed in terms of a finite number of functions of* $(e_{t-1}, e_{t-2}, \ldots)$. (Recall that, from (3.1.2), X_{t-u} is itself a function of $e_{t-1}, e_{t-2}, \ldots, u=1, \ldots, k$.) This idea takes us very close to Akaike's definition of a state vector (cf. Section 2.3), but we must remember that in the non-linear case we can no longer appeal to the linear space theory structure. This means, among other things, that we cannot now talk about a "minimal basis" (in the vector space sense). It is tempting to regard $\boldsymbol{x}_t = (e_{t-l+1}, \ldots, e_{t-1}, e_t, X_{t-k+1}, \ldots, X_{t-1}, X_t)'$ *as the "state vector at time t", and indeed this set of quantities may be so interpreted, but it does not follow that this vector leads to a minimal realization.* (In fact, we know that if the function h in (5.1.1) is linear, then a minimal realization of (5.1.1) has dimension $\max(k, l+1)$, whereas $\boldsymbol{x}_t$ has dimension $(k+l)$. Hence, in this case $\boldsymbol{x}_t$ does not provide a minimal realization unless $l=0$.)

It is worth remarking that even in the "system" case, where e_t is an observable input, the restriction of (3.1.1) to the form (5.1.1) would still be a very plausible one in many cases. What (5.1.1) says, in essence, is that the current "input", e_t, enters *linearly* in the expression for the current output, X_t, and this type of assumption is used extensively in the control theory literature on non-linear systems (see, for example, Brockett, 1976a).

The SDM formulation

Assuming that the function h in (5.1.1) is analytic, we may expand the RHS of (5.1.1) in a Taylor series about an arbitrary but fixed time point t_0. Using only a first-order expansion we obtain

$$\begin{aligned} X_t = h(X_{t_0-1}, \ldots, X_{t_0-k}, e_{t_0-1}, \ldots, e_{t_0-l}) \\ + \sum_{u=1}^{k} f_u(\boldsymbol{x}_{t-1})(X_{t-u} - X_{t_0-u}) \\ + \sum_{u=1}^{l} g_u(\boldsymbol{x}_{t-1})(e_{t-u} - e_{t_0-u}) + e_t, \quad \text{say}, \end{aligned} \tag{5.1.2}$$

where, as above, $\boldsymbol{x}_t$ denotes the "state vector",

$$\boldsymbol{x}_t = (e_{t-l+1}, \ldots, e_t, X_{t-k+1}, \ldots, X_t)',$$

and f_u, g_u depend on the first-order partial derivatives of h. We can now rewrite (5.1.2) in the form

$$X_t + \sum_{u=1}^{k} \phi_u(\boldsymbol{x}_{t-1})X_{t-u} = \mu(\boldsymbol{x}_{t-1}) + e_t + \sum_{u=1}^{l} \psi_u(\boldsymbol{x}_{t-1})e_{t-u}. \tag{5.1.3}$$

This is the basic model with which we will now work, and we call (5.1.3) a *State-dependent Model of order* (k, l). It has a natural and appealing interpretation as a locally linear ARMA model in which the evolution of the process at time $(t-1)$ is governed by a set of AR coefficients $\{\phi_u\}$, a set of MA coefficients $\{\psi_u\}$, and a local mean μ, all of which depend on the "state" of the process at time $(t-1)$.

By choosing particular forms for the $\{\phi_u\}$ and $\{\psi_u\}$, it is easily seen that the SDM (5.1.3) contains, as special cases, the (linear) ARMA models, the bilinear model, the threshold autoregressive, and the exponential autoregressive models. To obtain these special models we may proceed as follows:

(a) *Linear models.* Take μ, $\{\phi_u\}$, $\{\psi_u\}$ all as constants (i.e. independent of $\boldsymbol{x}_{t-1}$). Then (5.1.3) reduces to the usual ARMA (k, l) model.

(b) *Bilinear models.* Take μ, $\{\phi_u\}$, as constants, and set

$$\psi_u(\boldsymbol{x}_{t-1}) = c_u + \sum_{i=1}^{p} b_{iu} X_{t-i}; \qquad u = 1, \ldots, l. \tag{5.1.4}$$

Then (5.1.3) reduces to the bilinear model (4.1.4) (with the value of l in (5.1.3) chosen as $\max(r, k)$, r and k being the corresponding parameters in (4.1.1)). Note that for the bilinear model the $\psi_u(\boldsymbol{x}_{t-1})$ are constrained to be *linear functions* of $\{X_{t-1}, X_{t-2}, \ldots, X_{t-m}\}$.

(c) *Threshold AR models.* Take $\psi_u \equiv 0$ (all u), and

$$\mu_t = a_0^{(i)}$$

$$\phi_u(\boldsymbol{x}_{t-1}) = -a_u^{(i)}, \quad \text{if } X_{t-d} \in R^{(i)}. \tag{5.1.5}$$

Then (5.1.3) reduces to the threshold model, (4.2.7). Thus, for threshold models the ϕ_u are step-functions *which are constrained to depend on just one component of the state-vector*, namely X_{t-d}. The restrictive nature of threshold models is now apparent, and this may explain why threshold models seem best suited to cyclical data. (For cyclical data all one needs to know is whether one is on the "upper" or "lower" part of the cycle.) The more general type of threshold model, (4.2.4), gets closer to the spirit of the SDM approach, since here the ϕ_u depend on all components of the state vector.

(d) *Exponential AR models.* Take $\psi_u \equiv 0$ (all u), $\mu_t = 0$, and

$$\phi_u(\boldsymbol{x}_{t-1}) = -(\phi_u + \pi_u e^{-\gamma X_{t-1}^2}), \qquad u = 1, \ldots, k. \tag{5.1.6}$$

Then (5.1.3) reduces to the exponential AR model, (4.3.1). Note that for the exponential AR model the $\{\phi_u\}$ are exponential functions of just a single component X_{t-1}, rather like the feature noted above for threshold models.

State space representations of SDMs

We have already noted that, at time t, the future evolution of the SDM (5.1.3) is completely determined by the set of quantities,

$$\boldsymbol{x}_t = (e_{t-l+1}, \ldots, e_t, \vdots\, X_{t-k+1}, \ldots, X_t)',$$

together with the "innovation", e_{t+1}, and this set may thus be regarded as a "state vector" at time t. In fact, a formal state space description of (5.1.3) using the above "coordinates" is easily written down as follows:

$$\boldsymbol{x}_{t+1} = \boldsymbol{\mu}(\boldsymbol{x}_t) + \{\boldsymbol{F}(\boldsymbol{x}_t)\}\boldsymbol{x}_t + \boldsymbol{e}_{t+1}, \tag{5.1.7}$$

$$X_t = \boldsymbol{H}\boldsymbol{x}_t,$$

where

$$\boldsymbol{\mu}(\boldsymbol{x}_t) = (0, \ldots, 0 \,\vdots\, 0, \ldots, \mu(\boldsymbol{x}_t))' \tag{5.1.8}$$

$$\boldsymbol{e}_t = e_t(0, \ldots, 1 \,\vdots\, 0, \ldots, 1)' \tag{5.1.9}$$

$$\boldsymbol{H} = (0, \ldots, 0 \,\vdots\, 0, \ldots, 1), \tag{5.1.10}$$

and

$$\boldsymbol{F}(\boldsymbol{x}_t) = \left[\begin{array}{ccccc:cccccc} 0 & 1 & 0 & \cdots & 0 & 0 & \cdots & \cdots & \cdots & 0 \\ 0 & 0 & 1 & \cdots & 0 & 0 & \cdots & \cdots & \cdots & 0 \\ 0 & 0 & 0 & \cdots & 1 & 0 & \cdots & \cdots & \cdots & 0 \\ 0 & 0 & 0 & \cdots & 0 & 0 & \cdots & \cdots & \cdots & 0 \\ \hdashline 0 & 0 & \cdots & \cdots & 0 & 0 & 1 & 0 & \cdots & 0 \\ 0 & 0 & \cdots & \cdots & 0 & 0 & 0 & 1 & \cdots & 0 \\ 0 & 0 & \cdots & \cdots & 0 & 0 & 0 & 0 & \cdots & 1 \\ \psi_l & \psi_{l-1} & & & \psi_1 & -\phi_k & -\phi_{k-1} & & & -\phi_1 \end{array}\right]. \tag{5.1.11}$$

However, as we also noted previously, the representation (5.1.7) may well contain "redundant coordinates", i.e. the dimension of its state vector may be unecessarily large. If we assume that μ and the ϕ_u and ψ_u are analytic functions of $\boldsymbol{x}_t$, then (5.1.7) becomes a special case of a *linear-analytic* system, namely one of the form

$$\boldsymbol{x}_{t+1} = f\{\boldsymbol{x}_t\} + e_{t+1}\, g\{\boldsymbol{x}_t\} \tag{5.1.12}$$

$$X_t = k\{\boldsymbol{x}_t\}, \tag{5.1.13}$$

where f, g, and k are general analytic functions of their arguments. Such systems have been studied in considerable depth in the control theory literature (see, for example, Brockett, 1976a, 1976b, 1978) and, in particular, Sussman (1973, 1976) has generalized the notions of "controllability" and "observability" to the case of linear-analytic models. Sussman shows that if an input–output relationship admits a linear-analytic realization then there exists a "minimal" realization, namely one that is controllable and observable (in the above sense). Moreover, any two minimal realizations of the same input–output relationship are related by a "smooth" change of coordinates, and *all such realizations are based on state vectors of the same dimension.*

Thus, assuming that μ, $\{\phi_u\}$, $\{\psi_u\}$ are analytic, we may infer that there exists a realization of (5.1.3), of the form (5.1.7) in which $\boldsymbol{x}_t$ has a unique minimal dimension, n^* say. *We may now define n^* as the "dimension" of the SDM* (5.1.3). We note, in passing, that it is always possible to choose a new coordinate basis for a linear-analytic model such that X_t is a linear function of $\boldsymbol{x}_t$, i.e. so that (5.1.13) takes the form

$$X_t = \boldsymbol{H}\boldsymbol{x}_t,$$

$\boldsymbol{H}$ being a constant matrix (Brockett, 1976a).

Clearly, the value of n^* depends in quite a complicated way on the form of the functions μ, $\{\phi_u\}$, $\{\psi_u\}$. If all these parameters are constants (so that (5.1.3) is then a linear model), we know that $n^* = \max(k, l+1)$ and in this case (5.1.7) is clearly not a minimal realization unless $l = 0$ (i.e. unless the model is of the pure "AR" type). One may conjecture that n^* will be "fairly close" to the $\max(k, l+1)$ if μ, $\{\phi_u\}$, $\{\psi_u\}$, are all sufficiently "slowly varying" functions of $\boldsymbol{x}_t$, but without imposing specific assumptions on the behaviour of these functions it would not be possible to determine the value of n^* purely in terms of the parameters k and l. (Of course, if $l = 0$, i.e. if the model is of the pure "AR" type, then $n^* = k$, and the representation (5.1.7) is minimal.)

It is important to note, however, that the fact that the representation (5.1.7) may not be minimal does not in any way affect the identifiability of the SDM model (5.1.3). The model (5.1.3) is determined purely by the form of the function h in (5.1.1), and, *with the functional forms of* $\boldsymbol{\mu}$, $\boldsymbol{F}$, $\boldsymbol{H}$, *given in* (5.1.8)–(5.1.11), (5.1.7) may be regarded simply as a convenient matrix description of (5.1.3).

One particular feature which emerges from the above discussion is that we can no longer appeal to the "canonical forms" used in the state-space representation of linear models. For example, if we start from a representation of the form (cf. (2.3.5)–(2.3.9)),

$$\boldsymbol{x}_{t+1} = \{\boldsymbol{F}(\boldsymbol{x}_t)\}\boldsymbol{x}_t + \{\boldsymbol{G}(\boldsymbol{x}_t)\}e_{t+1},$$

$$X_t = \boldsymbol{H}\boldsymbol{x}_t,$$

with

$$F(\boldsymbol{x}_t) = \begin{bmatrix} 0 & 0 & \cdots & 0 & -\phi_n(\boldsymbol{x}_t) \\ 1 & 0 & \cdots & 0 & -\phi_{n-1}(\boldsymbol{x}_t) \\ \cdots & \cdots & \cdots & \cdots & \cdots \\ 0 & 0 & \cdots & 1 & -\phi_1(\boldsymbol{x}_t) \end{bmatrix},$$

$$G(\boldsymbol{x}_t) = [\psi_{n-1}(\boldsymbol{x}_t), \ldots, \psi_0(\boldsymbol{x}_t)]',$$

$$H = (0, 0, \ldots, 1),$$

then it is easily verified that this leads to the model,

$$X_t + \phi_1(\boldsymbol{x}_{t-1})X_{t-1} + \cdots + \phi_n(\boldsymbol{x}_{t-n})X_{t-n}$$
$$= \psi_0(\boldsymbol{x}_{t-1})e_t + \psi_1(\boldsymbol{x}_{t-2})e_{t-1} + \cdots + \psi_{n-1}(\boldsymbol{x}_{t-n})e_{t-n},$$

which is a form of SDM in which the "AR" coefficient, ϕ_u, depends on $\boldsymbol{x}_{t-u}$, and the "MA" coefficient, ψ_u, depends on $\boldsymbol{x}_{t-u-1}$.

Priestley (1980) describes an alternative method of deriving the SDM model (5.1.3) which is based on starting from a truncated Volterra series expansion of the form,

$$X_t = \sum_{u=0}^{l} g_u e_{t-u} + \sum_{u=0}^{l} \sum_{v=0}^{l} g_{uv} e_{t-u} e_{t-v} + \cdots. \tag{5.1.14}$$

Brockett (1976a) also considered linear analytic realizations of input-output relationships described by finite Volterra series, and showed that a finite dimensional linear-analytic realization exists if and only if the kernels of the Volterra series are "separable", i.e. if the coefficients of each of the terms in (3.1.3) are expressible in the form

$$g_{uvw\ldots} = \sum_{i=1}^{M} \gamma_i^{(1)}(u)\gamma_i^{(2)}(v)\gamma_i^{(3)}(w)\ldots,$$

M being a finite integer. Clearly, this condition is satisfied if each term in (3.1.3) involves only a finite summation (as in (5.1.14)), for then each kernel involves only a finite number of parameters. Thus, restricting (3.1.3) to the form (5.1.14) is a stronger condition than Brockett's condition of "separability", but both are aimed at the same objective, namely a way of expressing the assumption that the terms in the Volterra expansion are *finitely generated.* The basic assumption underlying the SDM, namely (5.1.1), is of a similar nature, but it expresses the "finiteness" of the input-output relationship in a more direct form.

Geometrical interpretation of SDMs

One of the characteristic features of linear models is that the mean-square one-step ahead predictor, $\tilde{X}_{t+1}$, belongs to the linear space, H_t, spanned by $X_t, X_{t-1}, X_{t-2}, \ldots$. More precisely, let $\mathscr{B}_t^{(X)} = \mathscr{B}_t(X_t, X_{t-1}, \ldots)$ be the sequence of Borel fields generated by $X_t, X_{t-1}, \ldots$. Then for each t, $\tilde{X}_{t+1} = E[X_{t+1} | \mathscr{B}_t^{(X)}] \in H_t$. This property does not, of course, hold for non-linear models, and in this case we may think of the space, P_t, generated by the $\{\tilde{X}_t\}$ as a "curved space". The basic assumption underlying the SDM is roughly equivalent to saying that the space P_t is *locally Euclidean*, but with a metric which changes with $\boldsymbol{x}_t$. The model (5.1.3) represents a *local linearization* of the general non-linear model (5.1.1), and, for each $\boldsymbol{x}_t$, (5.1.3) may be interpreted as representing the *tangent plane* to the surface $h(\boldsymbol{x}_t)$. In this sense we may regard the SDM as being formed by *bending the linear model* (a term due to A. F. M. Smith) around each point in the state-space.

These considerations now lead to the notion of a "state-dependent transfer function" associated with the SDM (5.1.3). For each $\boldsymbol{x}$, we define the "transfer function" at the point $\boldsymbol{x}$ as $h_{\boldsymbol{x}}(e^{-i\omega})$, where

$$h_{\boldsymbol{x}}(z) = \left\{1 + \sum_{u=1}^{l} \psi_u(\boldsymbol{x}) z^u\right\}\left\{1 + \sum_{u=1}^{k} \phi_u(\boldsymbol{x}) z^u\right\}^{-1}. \tag{5.1.15}$$

This "transfer function" describes the quasi-linear behaviour of the model in the neighbourhood of the point $\boldsymbol{x}$, and for fixed t_0 we can compute the "linearized output" as

$$\hat{X}_t = h_{\boldsymbol{x}_{t_0}}(B) \cdot e_t.$$

We may then use the expression

$$C(t, t_0) = E[|X_t - \hat{X}_t|^2]$$

as a measure of the *curvature* of the model in the neighbourhood of $\boldsymbol{x}_{t_0}$.

Non-linearity and non-stationarity

Since each of the coefficients, $\{\phi_u\}$, $\{\psi_u\}$, in (5.1.3) changes over time, it is possible to think of (5.1.3) as a linear model with time-dependent coefficients $\{\phi_u^{(t)}\}$, $\{\psi_u^{(t)}\}$. However, in this situation the "time-dependence" of the coefficients is tied down firmly to the state $\boldsymbol{x}_t$, i.e. if $\boldsymbol{x}_t \equiv \boldsymbol{x}_s$, then $\phi_u^{(t)} = \phi_u^{(s)}$, $\psi_u^{(t)} = \psi_u^{(s)}$, all u. On the other hand, if we allow $\phi_u^{(t)}$, $\psi_u^{(t)}$ to be arbitrary functions of t then (5.1.3) becomes a (linear) *non-stationary* model. A non-linear/non-stationary model would correspond to allowing the $\{\phi_u\}$ and $\{\psi_u\}$ to depend on both $\boldsymbol{x}_t$ and t, i.e. we would have $\phi_u = \phi_u(t, \boldsymbol{x}_t)$, $\psi_u = \psi_u(t, \boldsymbol{x}_t)$.

There is an interesting form of "duality" between the notions of non-stationarity and non-linearity. A natural way of dealing with non-stationarity models is to split up the parameter space (i.e. the time axis) into a large number of small segments, and regard the process as "locally stationary" within each segment. (A formal development of this approach leads to the description of non-stationary processes in terms of "evolutionary (time-dependent) spectra" to be discussed in Chapter 6.) Our treatment of non-linear models is based, in effect, on splitting up the *state-space* into a large number of small segments, and regarding the process as "locally linear" within each segment.

5.2 IDENTIFICATION OF SDMs

So far our discussion of SDMs has been entirely theoretical, and we now turn our attention to the problem of estimating, or "identifying", the functional form of the parameters $\mu(\boldsymbol{x}_t)$, $\phi_u(\boldsymbol{x}_t)$, $\psi_u(\boldsymbol{x}_t)$ from observational data.

The simplest non-trivial assumption we can make about these parameters is that each is a linear function of the state vector $\boldsymbol{x}_t$, so that, for example, we may write

$$\psi_u(\boldsymbol{x}_t) = \psi_u^{(0)} + \boldsymbol{x}_t' \cdot \boldsymbol{\beta}_u, \tag{5.2.1}$$

$$\phi_u(\boldsymbol{x}_t) = \phi_u^{(0)} + \boldsymbol{x}_t' \cdot \boldsymbol{\gamma}_u. \tag{5.2.2}$$

Whilst this form would be appropriate for bilinear models (with $\boldsymbol{\gamma}_u \equiv \boldsymbol{0}$, all u), it is hardly reasonable to suppose that all non-linear models can be approximated sufficiently accurately with the ψ_u, ϕ_u set as linear functions of $\boldsymbol{x}_t$. However, as long as the ψ_u, ϕ_u are "smooth" functions of $\boldsymbol{x}_t$, it would be quite reasonable to suppose that they may be represented *locally* as linear functions. This means that we can obtain a fairly general representation of these parameters by allowing both $\boldsymbol{\beta}_u$ and $\boldsymbol{\gamma}_u$ to be themselves state-dependent. If we do this we are then faced with the problem of specifying the functional form of $\boldsymbol{\beta}_u$ and $\boldsymbol{\gamma}_u$, but we can avoid this difficulty by simply letting $\boldsymbol{\beta}_u$ and $\boldsymbol{\gamma}_u$ "wander" over time, i.e. we allow $\boldsymbol{\beta}_u = \boldsymbol{\beta}_u^{(t)}$, and $\boldsymbol{\gamma}_u = \boldsymbol{\gamma}_u^{(t)}$ to depend purely on the time parameter. This endows the SDM with considerable flexibility, and enables us to incorporate some degree of non-stationarity (as well as non-linearity) into the model.

The basic strategy now is to allow $\boldsymbol{\beta}_u^{(t)}$ and $\boldsymbol{\gamma}_u^{(t)}$ to wander in the form of "random walks", and the estimation procedure would then determine, for each t, those values of $\boldsymbol{\beta}_u^{(t)}$ and $\boldsymbol{\gamma}_u^{(t)}$ which, roughly speaking, minimize the discrepancy between the observed value of X_{t+1} and its predictor, $\tilde{X}_{t+1}$, computed from the model. The estimation procedure is thus based on a sequential type of algorithm, similar in nature to the "Kalman filter" algorithm. We now give a more precise formulation of this approach.

First, to simplify the notation, we rewrite the state-space representation of the SDM in a slightly different form; we augment the state vector with the constant unity, i.e. we now write

$$\boldsymbol{x}_t = (1, e_{t-l+1}, \ldots, e_t; X_{t-k+1}, \ldots, X_t)'.$$

This device allows us to incorporate the term $\mu(\boldsymbol{x}_t)$ within the transition matrix, $\boldsymbol{F}(\boldsymbol{x}_t)$, the new form of $\boldsymbol{F}(\boldsymbol{x}_t)$ being obtained by bordering the matrix (5.1.1) with an additional row and column. Specifically, the new form is given by

$$\boldsymbol{F}(\boldsymbol{x}_t) =$$ 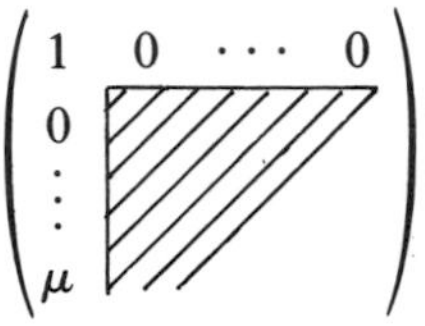 (5.2.3)

where the shaded portion is identical with (5.1.11), so that the final row becomes

$$(\mu \vdots \psi_l, \psi_{l-1}, \ldots, \psi_1 \vdots -\phi_k, -\phi_{k-1}, \ldots, -\phi_1).$$

(All these parameters are, of course, functions of $\boldsymbol{x}_t$.)

The SDM can now be written as

$$\boldsymbol{x}_{t+1} = \boldsymbol{F}(\boldsymbol{x}_t)\boldsymbol{x}_t + \boldsymbol{e}_{t+1}, \qquad (5.2.4)$$

$$X_t = \boldsymbol{H}\boldsymbol{x}_t,$$

where H is as given by (5.1.10), with the addition of an extra zero in the leading position, and $\boldsymbol{e}_t = e_t(0 \vdots 0 \cdots 1 \vdots 0 \cdots 1)$. We now denote the components of $\boldsymbol{x}_t$ by $(x_t^{(1)}, x_t^{(2)}, \ldots, x_t^{(m)})$, where $x_t^{(1)} \equiv 1$, $x_t^{(2)} = e_{t-l+1} \cdots$ $x_t^{(m)} = X_t$, and $m = k + l + 1$, and write

$$\Delta \boldsymbol{x}_t = (\Delta x_t^{(2)}, \Delta x_t^{(3)}, \ldots, \Delta x_t^{(m)})'. \qquad (5.2.5a)$$

(Note that $\Delta x_t^{(1)} \equiv 0$.) Then the above models for ψ_u, ϕ_u can be expressed in the form

$$\psi_u(\boldsymbol{x}_{t+1}) = \psi_u(\boldsymbol{x}_t) + \Delta \boldsymbol{x}'_{t+1}\boldsymbol{\beta}_u^{(t+1)}, \qquad (5.2.6)$$

$$\phi_u(\boldsymbol{x}_{t+1}) = \phi_u(\boldsymbol{x}_t) + \Delta \boldsymbol{x}'_{t+1}\boldsymbol{\gamma}_u^{(t+1)}, \qquad (5.2.7)$$

and we may adopt a similar model for $\mu(x_t)$, namely

$$\mu(\boldsymbol{x}_{t+1}) = \mu(\boldsymbol{x}_t) + \Delta \boldsymbol{x}'_{t+1}\boldsymbol{\alpha}^{(t+1)}, \quad \text{say.} \qquad (5.2.8)$$

The above recursions can be written in a more compact form by introducing the $(m-1)\times m$ matrix,

$$\Delta \boldsymbol{X}_t = (\boldsymbol{0}, \boldsymbol{0}, \ldots, \boldsymbol{0}, \Delta \boldsymbol{x}_t), \tag{5.2.9}$$

together with the $(m-1)\times m$ matrix,

$$\boldsymbol{B}_t = (\boldsymbol{\alpha}^{(t)} \vdots \boldsymbol{\beta}_l^{(t)}, \boldsymbol{\beta}_{l-1}^{(t)}, \ldots, \boldsymbol{\beta}_1^{(t)} \vdots \boldsymbol{\gamma}_k^{(t)}, \boldsymbol{\gamma}_{k-1}^{(t)}, \ldots, \boldsymbol{\gamma}_1^{(t)}). \tag{5.2.10}$$

Writing $\boldsymbol{F}_t \equiv F(x_t)$, the "updating" equations, (5.2.6), (5.2.7), and (5.2.8), can then be expressed in terms of the single equation,

$$\boldsymbol{F}_{t+1} = \boldsymbol{F}_t + \Delta \boldsymbol{X}'_{t+1} \boldsymbol{B}_{t+1}. \tag{5.2.11}$$

The random walk model for the "gradients", $\boldsymbol{\alpha}^{(t)}$, $\boldsymbol{\beta}_l^{(t)}, \ldots, \boldsymbol{\gamma}_1^{(t)}$, is now described by setting

$$\boldsymbol{B}_{t+1} = \boldsymbol{B}_t + \boldsymbol{V}_{t+1}, \tag{5.2.12}$$

where $\{\boldsymbol{V}_t\}$ is a sequence of independent matrix-valued random variables, the elements of each matrix having a multivariate normal distribution with zero means and variance-covariance matrix, $\boldsymbol{\Sigma}_v$, say. We write, symbolically, $\boldsymbol{V}_t = N(\boldsymbol{0}, \boldsymbol{\Sigma}_v)$. Finally, we assume that the innovations $\{e_t\}$ are independent normal variables, with zero mean and variance σ_e^2, so that $\boldsymbol{e}_t = N(\boldsymbol{0}, \boldsymbol{\Sigma}_e)$, where

$$\boldsymbol{\Sigma}_e = \sigma_e^2(0 \vdots \cdots 1 \vdots 0 \cdots 1)'(0 \vdots 0 \cdots 1 \vdots 0 \cdots 1).$$

To summarize the above, we now restate the complete recursion for the SDM.

$$X_t = \boldsymbol{H}\boldsymbol{x}_t, \tag{5.2.13}$$

$$\boldsymbol{x}_{t+1} = \boldsymbol{F}_t \boldsymbol{x}_t + \boldsymbol{e}_{t+1}, \tag{5.2.14}$$

$$\boldsymbol{F}_{t+1} = \boldsymbol{F}_t + \Delta \boldsymbol{X}'_{t+1} \boldsymbol{B}_{t+1}, \tag{5.2.15}$$

$$\boldsymbol{B}_{t+1} = \boldsymbol{B}_t + \boldsymbol{V}_{t+1}, \tag{5.2.16}$$

where $\boldsymbol{e}_t = N(0, \boldsymbol{\Sigma}_e)$, $\boldsymbol{V}_t = N(\boldsymbol{0}, \boldsymbol{\Sigma}_v)$.

In the form given above, there may seem to be a superficial resemblance between the SDM and the DLM ("dynamic linear model") scheme proposed by Harrison and Stevens (1976). There is, however, a fundamental distinction between the two types of models; in the DLM the parameters are arbitrary (random) functions of time, and, as its name implies, the DLM is a *linear* model with time-dependent coefficients. In the SDM scheme the parameters are tied down firmly to be *functions of the state* $\boldsymbol{x}_t$ (thus making the model non-linear), and this crucial feature is an essential part of the updating equation for $\boldsymbol{F}_t$.

Similar remarks may be made about the relationship between the SDM and the type of models considered by Young. In an important and extensive series of papers (see, for example, Young, 1970, 1974, 1975, 1978) he has developed various forms of recursive algorithms for estimating time-dependent parameters of linear systems, but, as in the case of Harrison and Stevens, this work is directed primarily at *linear* systems whose parameters are allowed to vary freely over time. However, in one of his later papers Young (1978) introduces in an informal way the notion that the time variation in the parameters of a model may be related to past values of the observed process—which is basically similar to the concept of "state-dependent" parameters. The same ideas are implicit in the comments by Young (1975).

Threshold models

We have seen that for threshold autoregressive models the $\{\phi_u\}$ are step-functions (cf. (5.1.5)), and thus the assumption that the $\{\phi_u\}$ are "smooth" functions is not strictly valid in this case. However, this does not present any serious difficulties since we can always approximate a step-function by a continuous function with large (but finite) gradients. More significantly, it may be argued that in practice the threshold model is essentially a device for describing a continuous non-linear relationship by a step-function approximation (cf. the approximation of (4.2.3) by (4.2.2)). Indeed, the physical examples of threshold models discussed by Tong (1983) are all of this type. The only case where the above feature might cause some difficulty is in the analysis of *artificial data* which are deliberately generated from a strict threshold model. We discuss the analysis of such data in the following section, and it will be seen that even in this case the SDM algorithm is still capable of providing a reasonably accurate identification of the original model.

Updating the $\boldsymbol{B}_t$*'s*

Since the parameters of an SDM are continually changing, the natural way to fit this type of model is to proceed sequentially. Let us suppose then that, on the basis of observations available up to time t, we have constructed an estimate, $\hat{\boldsymbol{B}}_{t-1}$, of $\boldsymbol{B}_{t-1}$, and an estimate, $\hat{\boldsymbol{F}}_{t-1}$, of $\boldsymbol{F}_{t-1}$. (Note that a model fitted at time t involves the transition matrix at time $(t-1)$, $\boldsymbol{F}_{t-1}$, (cf. (5.2.14), and hence estimates $\boldsymbol{B}_{t-1}$, $\boldsymbol{F}_{t-1}$, rather than $\boldsymbol{B}_t$, $\boldsymbol{F}_t$.) Before the next observation, X_{t+1}, becomes available, the obvious estimate of $\boldsymbol{B}_t$ is (from (5.2.16)),

$$\hat{\hat{\boldsymbol{B}}}_t = \hat{\boldsymbol{B}}_{t-1},$$

and, from (5.2.15), the corresponding estimate of $\boldsymbol{F}_t$ is

$$\hat{\hat{\boldsymbol{F}}}_t = \hat{\boldsymbol{F}}_{t-1} + \Delta \boldsymbol{X}'_t \hat{\boldsymbol{B}}_{t-1}.$$

The best predictor of X_{t+1} is now (from (5.2.13), (5.2.14)) given by

$$\begin{aligned}\tilde{X}_{t+1} &= \boldsymbol{H}\tilde{\boldsymbol{x}}_{t+1} = \boldsymbol{H}\hat{\hat{\boldsymbol{F}}}_t \boldsymbol{x}_t \\ &= \boldsymbol{H}\{\hat{\boldsymbol{F}}_{t-1} + \Delta \boldsymbol{X}'_t \hat{\boldsymbol{B}}_{t-1}\}\boldsymbol{x}_t.\end{aligned}$$

When the value of X_{t+1} becomes available we can compare this with the above predictor, $\hat{X}_{t+1}$, and then adjust the estimate $\hat{\hat{\boldsymbol{B}}}_t$ accordingly. The accuracy of $\hat{\hat{\boldsymbol{B}}}_t$ as an estimate of $\boldsymbol{B}_t$ will be reflected in the magnitude of the prediction error $(X_{t+1} - \tilde{X}_{t+1})$, and, guided by the form of the Kalman filter algorithm, one way to construct a revised estimate of $\boldsymbol{B}_t$ would be to take a weighted linear combination of $\hat{\hat{\boldsymbol{B}}}_t$ and $(X_{t+1} - \tilde{X}_{t+1})$. This leads to an updating algorithm of the form

$$\hat{\boldsymbol{B}}_t = \hat{\boldsymbol{B}}_{t-1} + \boldsymbol{K}_t[X_{t+1} - \boldsymbol{H}\{\hat{\boldsymbol{F}}_{t-1} + \Delta \boldsymbol{X}'_t \hat{\boldsymbol{B}}_{t-1}\}\boldsymbol{x}_t], \tag{5.2.17}$$

where the matrix $\boldsymbol{K}_t$ plays the role of the "Kalman gain" matrix. The corresponding updating equation for $\boldsymbol{F}_t$ is now

$$\hat{\boldsymbol{F}}_t = \hat{\boldsymbol{F}}_{t-1} + \Delta \boldsymbol{X}'_t \hat{\boldsymbol{B}}_t. \tag{5.2.18}$$

The idea previously mentioned that we may think of the SDM as being formed by *bending the linear model* is reinforced by the nature of the above algorithm for updating the $\boldsymbol{B}_t$ and $\boldsymbol{F}_t$ matrices. What we are doing, in effect, is taking a local linear model at time t and "bending" it so as to give the "best fit" to the next observation, X_{t+1}.

Ideally, we should choose $\boldsymbol{K}_t$ so as to minimize the "mean-square error", $E[\|\boldsymbol{B}_t - \hat{\hat{\boldsymbol{B}}}_t\|^2]$, but the non-linear form of the relationship between X_t and $\boldsymbol{B}_t$ would make it difficult to determine this optimal choice of $\boldsymbol{K}_t$. Although one could use (5.2.17) with an *ad hoc* choice of $\boldsymbol{K}_t$, a more satisfactory approach is to reformulate the SDM so that the complete set of quantities $\{\mu, \psi_1, \ldots, \psi_l, \phi_1, \ldots, \phi_k, \boldsymbol{\alpha}', \boldsymbol{\beta}'_1, \ldots, \boldsymbol{\beta}'_l, \boldsymbol{\gamma}'_1, \ldots, \boldsymbol{\gamma}'_k\}$ becomes the new "state". With this reformulation one can derive an explicit form of the optimal updating algorithms, based directly on the Kalman algorithm, and we discuss the details of this approach later in this section.

Young and Jakeman (1979) propose a recursive method of parameter estimation based on the "instrumental variable" technique which it is claimed has superior convergence properties and greater efficiency than the (extended) Kalman filter algorithm. However, in this approach the "instrumental variable" is taken as the hypothetical "noise-free" output of the system based on a preliminary estimate of the system's (rational) transfer function, and using the recorded values of the input process. Thus, this method would seem to be suitable only for the "systems" where one has an observable input.

Choice of $\boldsymbol{\Sigma}_v$

A suitable choice for the matrix $\boldsymbol{K}_t$ should clearly depend on $\boldsymbol{\Sigma}_v$, the variance–covariance matrix of the elements of $\boldsymbol{V}_t$. In fact, the relative magnitude of $\|\boldsymbol{\Sigma}_v\|$ to σ_e^2 should determine the "sensitivity" of the algorithm to changes in the $\boldsymbol{B}_t$. If $\|\boldsymbol{\Sigma}_v\|$ is small compared with σ_e^2 then the $\hat{\boldsymbol{B}}_t$ should not change very much but be effectively constant over time. (The algorithm will then attribute a large prediction error to the effect of the $\boldsymbol{e}_t$ rather than the $\boldsymbol{V}_t$.) On the other hand, if $\|\boldsymbol{\Sigma}_v\|$ is large compared with σ_e^2 then the $\hat{\boldsymbol{B}}_t$ should change rapidly in order to ensure a "good fit" for each new observation.

In practice we would start off the algorithm by fitting a conventional linear ARMA model to an initial stretch of the data. This would give initial estimates of μ, $\{\phi_u\}$, $\{\psi_u\}$, and σ_e^2 (together with appropriate values for k and l), and our choice of $\boldsymbol{\Sigma}_v$ then depends on how quickly we think the $\boldsymbol{B}_t$ are changing, i.e. $\boldsymbol{\Sigma}_v$ depends on our assumed "smoothness" of the $\boldsymbol{B}_t$.

Alternatively, we could make $\|\boldsymbol{\Sigma}_v\|$ time-dependent, and make it depend on the "distance" between $\boldsymbol{x}_t$ and $\boldsymbol{x}_{t-1}$. (If $\boldsymbol{x}_t$ is "close" to $\boldsymbol{x}_{t-1}$, then we would expect $\boldsymbol{B}_t$ to be "close" to $\boldsymbol{B}_{t-1}$.) This suggests that we might take $\|\boldsymbol{\Sigma}_v\| \propto \|\Delta \boldsymbol{x}_t\|^2$. Looking at it from a slightly different standpoint, we may argue that $\boldsymbol{V}_t$ is really the error involved in the local linearization of $\boldsymbol{F}(\boldsymbol{x}_t)$, and hence it would be sensible to make

$$\|\boldsymbol{\Sigma}_v\| \propto \Delta \boldsymbol{x}_t' \boldsymbol{F}''(\boldsymbol{x}_t) \Delta \boldsymbol{x}_t.$$

(The choice $\|\boldsymbol{\Sigma}_v\| \propto \|\Delta \boldsymbol{x}_t\|^2$ is, in effect, a rough approximation to the above form.)

Estimating the parameter "surfaces"

One of the main objectives in fitting SDMs is to obtain some idea of the non-linear structure of the process without first committing oneself to a particular form of non-linear model. To do this we need to construct estimates of the "surfaces" describing the functional form of the model parameters, $\{\psi_u(\boldsymbol{x}_t)\}$, $\{\phi_u(\boldsymbol{x}_t)\}$. We now describe a non-parametric method of estimating these "surfaces".

At time $(t+1)$ we obtain an estimate of $\boldsymbol{B}_t$, which in turn gives an estimate of $\boldsymbol{F}_t \equiv \boldsymbol{F}(\boldsymbol{x}_t)$. The final row of this matrix then gives us estimates, $\hat{\mu}(\boldsymbol{x}_t)$, $\hat{\psi}_1(\boldsymbol{x}_t), \ldots, \hat{\psi}_l(\boldsymbol{x}_t)$, $\hat{\phi}_1(\boldsymbol{x}_t), \ldots, \hat{\phi}_k(\boldsymbol{x}_t)$. We may now construct an $(m-1)$-dimensional plot of each of the parameters, plotted against the corresponding value of $\boldsymbol{x}_t$. As each new observation X_t becomes available we obtain a new ordinate in this graph, and as the observations continue we gradually build up a picture of each of the surfaces, $\hat{\mu}(\boldsymbol{x}_t)$, $\hat{\psi}_1(\boldsymbol{x}_t), \ldots, \hat{\phi}_k(\boldsymbol{x}_t)$.

More precisely, we may estimate each of the surfaces by a multi-dimensional form of the technique of *non-parametric function fitting* (see Priestley and Chao, 1972). Thus, suppose that we have a set of ordinates for the parameter ψ_1, say $\hat{\psi}_1(\boldsymbol{x}_1), \hat{\psi}_1(\boldsymbol{x}_2), \ldots, \hat{\psi}_1(\boldsymbol{x}_t)$. Then we may form a non-parametric estimate of the function $\psi_1(\boldsymbol{x})$ by computing

$$\hat{\psi}_1(\boldsymbol{x}) = \sum_{i=1}^{t} \hat{\psi}_1(\boldsymbol{x}_i) W(\boldsymbol{x}-\boldsymbol{x}_i) \prod_{k=2}^{m} (x^{(k)} - x_i^{(k)}), \tag{5.2.19}$$

where $W(\boldsymbol{x})$ is an $(m-1)$-dimensional "weight function". For example, we could take $W(\boldsymbol{x})$ as the density function of an $(m-1)$-dimensional $N(\mathbf{0}, \sigma_0^2 \boldsymbol{I})$ distribution, σ_0 then being a "bandwidth" parameter. Alternatively, we could take $W(\boldsymbol{x})$ to have a "rectangular" form, in which case $\hat{\psi}_1(\boldsymbol{x})$ is obtained simply by averaging neighbouring ordinates within a "grid" surrounding the point $\boldsymbol{x}$. (See Priestley and Chao (1972) for a discussion of various forms of weight functions appropriate to the one-dimensional case.)

Another possibility is to use a multi-dimensional spline-fitting procedure (see Wahba, 1975) on the ordinates, $\hat{\psi}_1(\boldsymbol{x}_1), \ldots, \hat{\psi}_1(\boldsymbol{x}_t)$. In fact, any multi-dimensional form of "smooth regression" technique may be used to construct the surface $\hat{\psi}_1(\boldsymbol{x})$.

If we use a "kernel" type estimate of the form (5.2.19), with $W(\boldsymbol{x})$ chosen as $N(\mathbf{0}, \sigma_0^2 \boldsymbol{I})$, say, then we have to specify the value of the "smoothing constant", σ_0. The choice of σ_0 interacts with the choice of Σ_v; if $\|\Sigma_v\|$ is large, the $\boldsymbol{B}_t$, and the $\boldsymbol{F}_t$, will be erratic, and hence we may use a large σ_0 in order to produce heavy smoothing. On the other hand, if $\|\Sigma_v\|$ is small, the $\boldsymbol{B}_t$ and $\boldsymbol{F}_t$ will be smooth, and little smoothing will be required from the kernel $W(\boldsymbol{x})$.

At each new observation, X_t, is obtained, a new estimate, $\hat{\psi}_1(\boldsymbol{x}_t)$, is formed, and we have a new "point" to insert in the graph—leading to an improved estimate of the surface, $\psi_1(\boldsymbol{x})$. Thus, as observations on X_t are accumulated we continually build up and improve the non-parametric estimate of each of the surfaces, $\mu(\boldsymbol{x})$; $\psi_1(\boldsymbol{x}), \ldots, \psi_l(\boldsymbol{x})$; $\phi_1(\boldsymbol{x}), \ldots, \phi_k(\boldsymbol{x})$.

The above technique presupposes that at time $(t+1)$ each element of the state vector, $\boldsymbol{x}_t$, is known. In the case of the pure "AR" type of SDM the elements of $\boldsymbol{x}_t$ consist simply of past observations, $X_t, X_{t-1}, \ldots, X_{t-k+1}$, and these are, of course, all known. In the "MA" or "ARMA" type of SDM some of the elements of x_t consist of past innovations, $e_t, e_{t-1}, \ldots, e_{t-l+1}$. If we have a sufficiently long sequence of observations on X_t we may determine the (fitted) innovations recursively from (5.1.3) (using suitable starting values). Alternatively, we may estimate the full state vector, $\boldsymbol{x}_t$, at each time point by applying the "extended Kalman filter" algorithms directly to (5.2.13), (5.2.14), using, for each t, the fitted form of $\boldsymbol{F}_t$.

In practice it would probably suffice to assume that the parameters μ, $\psi_1, \ldots, \psi_l$; $\phi_1, \ldots, \phi_k$, are functions of $(X_t, X_{t-1}, \ldots, X_{t-k+1})$ only (even for "ARMA" type SDMs). Even with this restriction the resulting SDM is still flexible enough to cover linear, bilinear, threshold autoregressive, and exponential autoregressive models, and a plot of the $\hat{\psi}_u$, $\hat{\phi}_u$ as functions purely of $(X_t, X_{t-1}, \ldots, X_{t-k+1})$ would probably be sufficient to indicate the general non-linear structure of the model.

It is important to note that for specific non-linear models, such as bilinear, threshold AR, and exponential AR, the fitted parameters, $\hat{\psi}_1, \ldots, \hat{\psi}_l$; $\hat{\phi}_1, \ldots, \hat{\phi}_k$, will not necessarily reproduce the "natural" functional forms given by (5.1.4), (5.1.5), (5.1.6). The essential point is that the fitting algorithm described above fits, for each t, a *local* linear model. In terms of the general model, (5.1.5), this means, in effect, that the $\hat{\psi}_u$ and $\hat{\phi}_u$ will follow the pattern of *the first partial derivatives of h w.r.t. e_{t-u}, X_{t-u}, respectively*. For the threshold autoregressive AR model (4.2.7) we have

$$h(X_{t-1}, \ldots, X_{t-k}; e_{t-1}, \ldots, e_{t-l}) = a_0^{(i)} + \sum_{u=1}^{k} a_u^{(i)} X_{t-u}, \qquad \text{if } X_{t-d} \in R^{(i)}$$

so that

$$\hat{\phi}_u \sim -\frac{\partial h}{\partial X_{t-u}} = -a_u^{(i)}, \qquad \text{if } X_{t-d} \in R^{(i)}.$$

Thus, in this case the $\{\hat{\phi}_u\}$ will follow the "natural" form. However, for the exponential AR model,

$$h(X_{t-1}, \ldots, X_{t-k}; e_{t-1}, \ldots, e_{t-l}) = \sum_{u=1}^{k} (\phi_u + \pi_u e^{-\gamma X_{t-1}^2}) X_{t-u}.$$

Hence

$$\hat{\phi}_u \sim -\frac{\partial h}{\partial X_{t-u}} = -(\phi_u + \pi_u e^{-\gamma X_{t-1}^2}); \qquad u = 2, 3, \ldots, k,$$

but

$$\hat{\phi}_1 \sim -\frac{\partial h}{\partial X_{t-1}} = -(\phi_1 + \pi_1 e^{-\gamma X_{t-1}^2}) + 2\gamma X_{t-1} e^{-\gamma X_{t-1}^2} \left[\sum_{j=1}^{k} \pi_j X_{t-j} \right].$$

For the bilinear model

$$h(X_{t-1}, \ldots, X_{t-k}; e_{t-1}, \ldots, e_{t-l}) = -\sum_{u=1}^{p} a_u X_{t-u} + \sum_{u=1}^{r} c_u e_{t-u} + \sum_{i=1}^{m} \sum_{j=1}^{k} b_{ij} e_{t-j} X_{t-i}.$$

Here,

$$\hat{\phi}_u \sim -\frac{\partial h}{\partial X_{t-u}} = a_u + \sum_{j=1}^{k} b_{uj} e_{t-j},$$

$$\hat{\psi}_u \sim \frac{\partial h}{\partial e_{t-u}} = c_u + \sum_{i=1}^{m} b_{iu} X_{t-i}.$$

The $\hat{\psi}_u$ follow the form given by (5.1.4), but the $\hat{\phi}_u$ contain an additional term, linear in the $\{e_t\}$.

If the $\{\hat{\psi}_u\}$, $\{\hat{\phi}_u\}$ surfaces give strong indications that one of the above models is, in fact, appropriate, then we would go back to the data and fit the specific model indicated, using efficient parameter estimation procedures. However, the $\{\hat{\psi}_u\}$, $\{\hat{\phi}_u\}$ surfaces would still give useful information in the form of rough initial estimates for the model parameters. For example, if we decide that a threshold AR model is suitable, the $\{\psi_u\}$, $\{\phi_u\}$ surfaces might give us a prior indication of the number of thresholds, their approximate locations, and on which of the $X_{t-1}, \ldots, X_{t-k}$ they are most strongly dependent.

Forecasting with SDMs

Apart from its use in indicating a particular type of non-linear structure, the SDM may be used directly as a forecasting model. Thus, if we are given observations on $\{X_t\}$ up to time t, we may forecast X_{t+h} from

$$\tilde{X}_{t+h} = \boldsymbol{H}\hat{\boldsymbol{x}}_{t+h}, \tag{5.2.20}$$

where $\hat{\boldsymbol{x}}_{t+h}$ is computed recursively from (using (5.2.14)),

$$\hat{\boldsymbol{x}}_{t+j} = \hat{\boldsymbol{F}}_{t+j-1}\hat{\boldsymbol{x}}_{t+j-1}; \qquad j = 1, 2, \ldots, h, \tag{5.2.21}$$

and $\boldsymbol{F}_{t+j-1}$ is computed from (using (5.2.15)),

$$\hat{\boldsymbol{F}}_{t+j-1} = \hat{\boldsymbol{F}}_{t+j-2} + \Delta\hat{\boldsymbol{X}}'_{t+j-1}\hat{\boldsymbol{B}}_{t+j-1}; \qquad j = 1, 2, \ldots, h. \tag{5.2.22}$$

Of course, if our observations extend only up to time t, the $\hat{\boldsymbol{B}}$ matrices are known only up to time $(t-1)$. For short-term forecasting (h small) we may set

$$\hat{\boldsymbol{B}}_{t+j-1} \equiv \hat{\boldsymbol{B}}_{t-1}, \qquad \text{all } j, \tag{5.2.23}$$

i.e. we set all the $\hat{\boldsymbol{B}}_{t+j-1}$ equal to the last available estimate $\hat{\boldsymbol{B}}_{t-1}$. Under (5.2.23), the forecasts are based on the assumption that, between times t and $(t+h)$, μ, $\{\psi_u\}$, $\{\phi_u\}$, are strictly *linear* functions of $\boldsymbol{x}_t$.

However, for longer-term forecasting the assumption of strict linearity in the coefficients μ, $\{\psi_u\}$, $\{\phi_u\}$, would be undesirable, and in this case it would be preferable to estimate each $\hat{\boldsymbol{F}}_{t+j-1}$ by evaluating $\hat{\mu}(\boldsymbol{x}_{t+j-1})$, $\{\hat{\psi}_u(\boldsymbol{x}_{t+j-1})\}$, $\{\hat{\phi}_u(\boldsymbol{x}_{t+j-1})\}$, from the "surfaces" fitted up to time t, but evaluated at the point $\hat{\boldsymbol{x}}_{t+j-1}$.

Thus, if we denote the parameter surfaces fitted up to time to be $\hat{\mu}^{(t)}(x)$, $\hat{\psi}_u^{(t)}(\boldsymbol{x})$, $\hat{\phi}_u^{(t)}(\boldsymbol{x})$, then the elements of the final row of $\hat{\boldsymbol{F}}_{t+j-1}$ are computed as

$$\hat{\mu} = \hat{\mu}^{(t)}(\hat{\boldsymbol{x}}_{t+j-1})$$

$$\hat{\psi}_u = \hat{\psi}_u^{(t)}(\hat{\boldsymbol{x}}_{t+j-1}), \qquad u = 1, \ldots, l;$$

$$\hat{\phi}_u = \hat{\phi}_u^{(t)}(\hat{\boldsymbol{x}}_{t+j-1}), \qquad u = 1, \ldots, k.$$

(The $\hat{\boldsymbol{x}}_{t+j-1}$ are computed as in (5.2.21).) With this approach the forecasts are no longer based on parameters constrained to be linear functions of $\boldsymbol{x}_t$. In particular, the "eventual forecast function" generated from (5.2.21) could well give rise to *limit cycle behaviour*, provided the fitted parameter surfaces so indicate, i.e. provided the past data tell us that limit cycle behaviour is appropriate.

"AR" type SDMs

As in the case of linear models, the estimation procedures are considerably simplified when the model is of the pure "AR" type. An "AR" SDM takes the form

$$X_t = \mu_{t-1} - \phi_{1,t-1}X_{t-1} - \phi_{2,t-1}X_{t-2} - \cdots - \phi_{k,t-1}X_{t-k} + e_t, \quad (5.2.24)$$

where (cf. (5.2.8)),

$$\begin{aligned}\mu_t &= \mu_{t-1} + \alpha_1^{(t)}\Delta x_t^{(2)} + \alpha_2^{(t)}\Delta_2^{(t)}\Delta x_t^{(3)} + \cdots + \alpha_k^{(t)}\Delta x_t^{(k+1)} \\ &= \mu_{t-1} + \Delta \boldsymbol{x}_t' \cdot \boldsymbol{\alpha}^{(t)}, \end{aligned} \quad (5.2.25)$$

and for each u (cf. (5.2.7)),

$$\begin{aligned}\phi_{u,t} &= \phi_{u,t-1} + \gamma_{u1}^{(t)}\Delta x_t^{(2)} + \gamma_{u2}^{(t)}\Delta x_t^{(3)} + \cdots + \gamma_{uk}^{(t)}\Delta x_t^{(k+1)} \\ &= \phi_{u,t-1} + \Delta \boldsymbol{x}_t' \boldsymbol{\gamma}_u^{(t)}. \end{aligned} \quad (5.2.26)$$

Also,

$$(\boldsymbol{\alpha}^{(t)} \vdots \boldsymbol{\gamma}_k^{(t)} \ldots \boldsymbol{\gamma}_1^{(t)}) = (\boldsymbol{\alpha}^{(t-1)} \vdots \boldsymbol{\gamma}_k^{(t-1)} \ldots \boldsymbol{\gamma}_1^{(t-1)}) + \boldsymbol{V}_t. \quad (5.2.27)$$

We can now rewrite (5.2.24) in the form

$$X_t = \boldsymbol{H}_t^* \boldsymbol{\theta}_t + e_t, \quad (5.2.28)$$

where $\boldsymbol{H}_t^*$ is a $1 \times (k+1)^2$ row vector given by,

$$\boldsymbol{H}_t^* = (1, -X_{t-1}, \ldots, -X_{t-k}; 0, 0, \ldots, 0), \quad (5.2.29)$$

(note that there are $k(k+1)$ 0s in $\boldsymbol{H}_t^*$), and $\boldsymbol{\theta}_t$ is a $(k+1)^2 \times 1$ column vector given by,

$$\boldsymbol{\theta}_t = (\mu_{t-1}, \phi_{1,t-1}, \ldots, \phi_{k,t-1} \vdots \boldsymbol{\alpha}^{(t)'}, \boldsymbol{\gamma}_1^{(t)'}, \ldots, \boldsymbol{\gamma}_k^{(t)'})'. \tag{5.2.30}$$

The evolution of $\boldsymbol{\theta}_t$ over time is given by

$$\boldsymbol{\theta}_t = \boldsymbol{F}_{t-1}^* \boldsymbol{\theta}_{t-1} + \boldsymbol{W}_t, \tag{5.2.31}$$

where the $(k+1)^2 \times (k+1)^2$ matrix $\boldsymbol{F}_{t-1}^*$ is given by

$$\boldsymbol{F}_{t-1}^* = \left[\begin{array}{c|ccc} & \Delta\boldsymbol{x}_{t-1}' & & \boldsymbol{0} \\ & & \Delta\boldsymbol{x}_{t-1}' & \\ \boldsymbol{I}_{k+1} & & & \\ & \boldsymbol{0} & & \Delta\boldsymbol{x}_{t-1}' \\ \hline & & & \\ \boldsymbol{0} & & \boldsymbol{I}_{k(k+1)} & \end{array}\right] \begin{array}{c} \uparrow \\ (k+1) \text{ rows} \\ \downarrow \\ \cdots \\ \uparrow \\ k(k+1) \text{ rows} \\ \downarrow \end{array} \tag{5.2.31a}$$

and

$$\boldsymbol{W}_t = (0, 0, \ldots, 0; \boldsymbol{v}_{1,t}', \ldots, \boldsymbol{v}_{k+1,t}')', \tag{5.2.32}$$

$\boldsymbol{v}_{1,t}, \ldots, \boldsymbol{v}_{k+1,t}$ being the columns of $\boldsymbol{V}_t$. (Note that $\boldsymbol{W}_t$ contains $(k+1)$ 0 elements, and that here $\Delta\boldsymbol{x}_{t-1}' = (X_{t-1} - X_{t-2}, X_{t-2} - X_{t-3}, \ldots, X_{t-k} - X_{t-k-1})$.) Equations (5.2.28) and (5.2.31) now have the standard form to which the (extended) Kalman filter algorithm can be applied immediately. In fact, (5.2.28) and (5.2.31) assume a form very similar to the Harrison–Stevens model (Harrison and Stevens, 1976), *with the important distinction that here the* $\boldsymbol{F}_{t-1}$ *matrix is a function of* $\Delta\boldsymbol{X}_{t-1}$. If we think of $\boldsymbol{\theta}_t$ as the "state" and $\boldsymbol{H}_t^*$ as the "observation" matrix, then a direct application of the Kalman algorithm to (5.2.28) and (5.2.31) gives the recursion (see, for example, Priestley, 1978a).

$$\hat{\boldsymbol{\theta}}_t = \boldsymbol{F}_{t-1}^* \hat{\boldsymbol{\theta}}_{t-1} + \boldsymbol{K}_t^* \{X_t - \boldsymbol{H}_t^* \boldsymbol{F}_{t-1}^* \hat{\boldsymbol{\theta}}_{t-1}\}, \tag{5.2.33}$$

where $\boldsymbol{K}_t^*$, the Kalman "gain" matrix, is given by

$$\boldsymbol{K}_t^* = \boldsymbol{\Phi}(\boldsymbol{H}_t^*)' \sigma_{\tilde{e}}^{-2} \tag{5.2.34}$$

$\boldsymbol{\Phi}$ being the variance–covariance matrix of the one-step prediction errors of $\boldsymbol{\theta}_t$, i.e.

$$\boldsymbol{\Phi} = E[\{\boldsymbol{\theta}_t - \boldsymbol{F}_{t-1}^* \hat{\boldsymbol{\theta}}_{t-1}\}\{\boldsymbol{\theta}_t - \boldsymbol{F}_{t-1}^* \hat{\boldsymbol{\theta}}_{t-1}\}'],$$

and $\sigma_{\tilde{e}}^2$ the variance of the one-step prediction errors of X_t, i.e. $\sigma_{\tilde{e}}^2$ is the variance of $\tilde{e}_t = \{X_t - \boldsymbol{H}_t^* \boldsymbol{F}_{t-1}^* \hat{\boldsymbol{\theta}}_{t-1}\}$.

Equation (5.2.33) gives an explicit form of the "updating" algorithm for the parameters $\mu_t, \phi_{1,t}, \ldots, \phi_{k,t}$, similar in nature to (5.2.17) and (5.2.18), but here we can exploit the fact that (5.2.28) and (5.2.31) are of "standard form", and hence we can use the standard expression for the Kalman gain matrix $\boldsymbol{K}_t^*$.

Noting that $\tilde{e}_t = \boldsymbol{H}_t^*(\boldsymbol{\theta}_t - \boldsymbol{F}_{t-1}^*\hat{\boldsymbol{\theta}}_{t-1}) + e_t$, then $\sigma_{\tilde{e}}^2 = \{\boldsymbol{H}_t^*\boldsymbol{\Phi}_t(\boldsymbol{H}_t^*)' + \sigma_e^2\}$. If we now denote the variance-covariance matrix of $(\boldsymbol{\theta}_t - \hat{\boldsymbol{\theta}}_t)$ by $\boldsymbol{C}_t$, i.e. if we set $\boldsymbol{C}_t = E[(\boldsymbol{\theta}_t - \hat{\boldsymbol{\theta}}_t)(\boldsymbol{\theta}_t - \hat{\boldsymbol{\theta}}_t)']$, then we may use the standard recursive equations for the Kalman filter, namely (see, for example, *Spec. Anal.*, p. 810),

$$\boldsymbol{K}_t^* = \boldsymbol{\Phi}_t(\boldsymbol{H}_t^*)'[\boldsymbol{H}_t^*\boldsymbol{\Phi}_t(\boldsymbol{H}_t^*)' + \sigma_e^2]^{-1},$$

$$\boldsymbol{\Phi}_t = \boldsymbol{F}_{t-1}^*\boldsymbol{C}_{t-1}(\boldsymbol{F}_{t-1}^*)' + \boldsymbol{\Sigma}_w,$$

$$\boldsymbol{C}_t = \boldsymbol{\Phi}_t - \boldsymbol{K}_t^*[\boldsymbol{H}_t^*\boldsymbol{\Phi}_t(\boldsymbol{H}_t^*)' + \sigma_e^2](\boldsymbol{K}_t^*)',$$

where

$$\boldsymbol{\Sigma}_w = \begin{bmatrix} \mathbf{0} & \mathbf{0} \\ \mathbf{0} & \boldsymbol{\Sigma}_v \end{bmatrix}.$$

Choice of $\boldsymbol{\Sigma}_w$

The matrix $\boldsymbol{\Sigma}_w$ depends effectively on $\boldsymbol{\Sigma}_v$, the variance-covariance matrix of the elements of $\boldsymbol{V}_t$. As noted previously, the choice of $\boldsymbol{\Sigma}_v$ depends on our assumed "smoothness" of the model parameters as functions of $\boldsymbol{x}_t$. In the exploratory stage of the analysis it would be wise to choose $\boldsymbol{\Sigma}_v$ so that its diagonal elements are all large relative to $\hat{\hat{\sigma}}_e^2$; this would then give the updating algorithm freedom to make fairly rapid changes in the model parameters. If, as a result, the parameters turn out to be highly erratic as functions of $\boldsymbol{x}_t$, this can be corrected at a later stage by an appropriate choice of smoothing parameters when constructing the "parameter surfaces".

Extension to general SDM schemes

The above approach is readily extended to the more general "ARMA" type SDM by redefining $\boldsymbol{H}_t^*$ and $\boldsymbol{\theta}_t^*$ as follows: set

$$\boldsymbol{H}_t^* = (1, e_{t-1}, \ldots, e_{t-l}, -X_{t-1}, \ldots, -X_{t-k}, 0, 0, \ldots, 0),$$

$$\boldsymbol{\theta}_t = (\mu_{t-1}, \psi_{1,t-1}, \ldots, \psi_{l,t-1}, \phi_{1,t-1}, \ldots, \phi_{k,t-1}, \boldsymbol{\alpha}^{(t)'}, \boldsymbol{\beta}_1^{(t)'}, \ldots, \boldsymbol{\beta}_l^{(t)'}, \boldsymbol{\gamma}_1^{(t)'}, \ldots, \boldsymbol{\gamma}_k^{(t)'})'.$$

The matrix $\boldsymbol{F}^*_{t-1}$ retains the form (5.3.31a) (but with Δx_{t-1} given by the more general expression (5.2.5a)) and $\boldsymbol{W}_t$ the form (5.2.32), except that now k is replaced by $(k+l)$, so that, for example, $\boldsymbol{F}^*_{t-1}$ is now a $(k+l+1)^2 \times (k+l+1)^2$ matrix. The optimal updating algorithm for $\hat{\boldsymbol{\theta}}_t$ is now given by (5.2.33), with $\boldsymbol{K}^*_t$ given by (5.2.34). However, with this formulation, we need to know both the matrices $\boldsymbol{F}^*_{t-1}$ and $\boldsymbol{H}^*_t$ at time t. This means that we need to know $e_{t-1}, \ldots, e_{t-l}$ in addition to $X_{t-1}, \ldots, X_{t-k}$. Given a sufficiently long sequence of observations on X_t, the relevant e_ts can be estimated recursively by setting, for each t,

$$\hat{e}_t = (X_t - \boldsymbol{H}^*_t \hat{\boldsymbol{\theta}}_t),$$

i.e. $\hat{e}_t$ is computed as the residual from the model fitted at time t. We now replace $e_{t-1}, \ldots, e_{t-l}$ in $\boldsymbol{H}^*_t$ by $\hat{e}_{t-1}, \ldots, \hat{e}_{t-l}$, and replace $\Delta e_{t-l}, \ldots, \Delta e_{t-1}$ by $\Delta \hat{e}_{t-l}, \ldots, \Delta \hat{e}_{t-1}$ in the vector $\Delta \boldsymbol{x}'_{t-1}$.

Starting values

In practice, the recursive procedure described above must be started at some value of $t = t_0$, say, and hence initial values are required for $\hat{\boldsymbol{\theta}}_{t_0-1}$ and $\boldsymbol{C}_{t_0-1}$. In finding the initial values use may be made of the fact that (5.1.3) represents a "locally" linear ARMA model, and hence we may formulate a practical estimation procedure as follows.

1. Taken an initial stretch of the data, say the first $2m$ observations, and fit an ARMA(k, l) model. This will provide initial values $\hat{\hat{\mu}}, \hat{\hat{\psi}}_1, \ldots, \hat{\hat{\psi}}_l, \hat{\hat{\psi}}_1, \ldots, \hat{\hat{\phi}}_k$ and $\hat{\hat{\sigma}}^2_e$, the residual variance of the model. An approximate method of finding initial estimates of the coefficients of an ARMA(k, l) model is to first fit a large-order AR model to the data, calculate the estimated residuals $\{\hat{e}_t\}$, and then fit a linear transfer function model between $\{X_t\}$ and $\{\hat{e}_t\}$. Since the estimates $\hat{\boldsymbol{\theta}}_{t_0-1}$, $\boldsymbol{C}_{t_0-1}$ are only to be used as initial values in the recursion procedure, this method of estimation is sufficiently accurate, and much more economical in computer time and storage than a more accurate method involving non-linear optimization.

2. The recursion is started midway along the initial stretch of data at $t_0 = m$ (where it seems reasonable to assume that the initially estimated parameter values are most accurate), setting

$$\hat{\boldsymbol{\theta}}_{t_0-1} = (\hat{\hat{\mu}}, \hat{\hat{\psi}}_1, \ldots, \hat{\hat{\psi}}_l, \hat{\hat{\phi}}_1, \ldots, \hat{\hat{\phi}}_k, 0, \ldots, 0)'$$

and

$$\boldsymbol{C}_{t_0-1} = \begin{bmatrix} \hat{\hat{\boldsymbol{R}}}_{\mu,\psi,\phi} & 0 \\ 0 & 0 \end{bmatrix}$$

where $\hat{\hat{\boldsymbol{R}}}_{\mu,\psi,\phi}$ is the estimated variance-covariance matrix of $(\hat{\hat{\mu}}, \hat{\hat{\psi}}_1, \ldots, \hat{\hat{\psi}}_l, \hat{\hat{\phi}}_1, \ldots, \hat{\hat{\phi}}_k)$ obtained from the initial ARMA(k, l) model fitting. If it is assumed that the initial values are reasonably accurate at $t_0 = m$, it also seems reasonable to set all the initial gradients to zero. It remains to choose reasonable values for $\boldsymbol{\Sigma}_v$, the variance-covaraince matrix of $\boldsymbol{V}_t$, (and hence, by implication, to choose values for $\boldsymbol{\Sigma}_w$). As noted previously, the choice of $\boldsymbol{\Sigma}_v$ depends on the assumed "smoothness" of the model parameters as functions of $\boldsymbol{x}_t$. In practice, one would wish initially to have the diagonal elements of $\boldsymbol{\Sigma}_v$ reasonably large relative to $\hat{\hat{\sigma}}_e^2$, to give the updating algorithm freedom to make fairly rapid changes in the model parameters. The diagonal elements of $\boldsymbol{\Sigma}_v$ are set equal to $\hat{\hat{\sigma}}_e^2$ multiplied by *some constant* α *called the "smoothing factor,"* and the off-diagonal elements are set equal to zero. However, if the elements of $\boldsymbol{\Sigma}_v$ are set too large, the estimated parameters become very unstable indeed, and tend to "explode" to very large values. This is clearly an unsatisfactory situation, but if the elements of $\boldsymbol{\Sigma}_v$ are made too small it is difficult to detect the non-linearity present in the data, since the procedure is then virtually equivalent to the recursive fitting of a linear model. The best procedure in practice appears to be to reduce the magnitude of the "smoothing factor" until the parameters exhibit non-explosive behaviour. If the parameters still appear to be far from "smooth", the smoothing factor may be reduced further. In addition, the parameters may be smoothed by a multi-dimensional form of the non-parametric function fitting technique of Priestley and Chao (1972). Having carried out this procedure, the resulting parameter surfaces should then give a clearer idea of the type of non-linearity present in the model, and will provide indications of the special type of non-linear model which should be fitted to the data. However, the multi-dimensional surface fitting technique poses problems which have yet to be resolved and so in this study the numerical examples use curve fitting only.

Generalizations

The SDM scheme may be generalized in a number of ways. Here we mention two possible extensions.

1. We have assumed that the parameters $\mu, \phi_1, \ldots, \phi_l, \psi_1, \ldots, \psi_k$, could be expressed as local *linear* functions of $\boldsymbol{x}_t$. However, in certain applications there may be strong physical reasons for choosing some other functional form of $\boldsymbol{x}_t$. For example, in an electrical circuit (such as an audio amplifier) the form of the "transfer function" between input and output may vary with the power level of the input. (It is well known that non-linear "distortion" effects in solid-state amplifiers take different forms depending on whether the input is a low power level or high power level.) Now at time

t a rough measure of the "local" input power level is

$$\frac{1}{l}\sum_{u=1}^{l} e_{t-u}^2 \propto \boldsymbol{e}'_{t-1}\boldsymbol{e}_{t-1},$$

where $\boldsymbol{e}'_t = (e_t, e_{t-1}, \ldots, e_{t-l+1})$. Hence, in this case it would be sensible to modify the updating equation for, e.g., ψ_u (cf. (5.2.6)) to

$$\psi_u^{(t+1)} = \psi_u^{(t)} + \Delta\boldsymbol{x}'_t\boldsymbol{\beta}_{1,u}^{(t+1)} + \Delta(\boldsymbol{x}'_t\boldsymbol{x}_t)\boldsymbol{\beta}_{2,u}^{(t+1)} \tag{5.2.35}$$

thus allowing ψ_u to be a locally quadratic function of $\boldsymbol{x}_t$.

2. We may modify the updating equations for $\boldsymbol{F}_t$ (cf. (5.2.11)) by introducing a "transition matrix" $\boldsymbol{G}_t$ and a further white noise process $\boldsymbol{Z}_t$ so that (5.2.11) becomes

$$\boldsymbol{F}_{t+1} = \boldsymbol{G}_t\boldsymbol{F}_t + \Delta\boldsymbol{X}'_t\boldsymbol{B}_{t+1} + \boldsymbol{Z}_{t+1}. \tag{5.2.36}$$

The extra term, $\boldsymbol{Z}_{t+1}$, injects some random time variations into the model parameters (in addition to their dependence on the state variable, $\boldsymbol{x}_t$), and if we form the parameters into a vector $\boldsymbol{\theta}_t$, and reformulate the model as above we obtain an equation of the form

$$\boldsymbol{\theta}_t = \boldsymbol{G}^*_{t-1}\boldsymbol{\theta}_{t-1} + \boldsymbol{W}^*_t, \tag{5.2.37}$$

where

$$\boldsymbol{W}^*_t = (\boldsymbol{Z}_t \vdots\, \boldsymbol{v}_{1,t+1}, \ldots, \boldsymbol{v}_{k+1,t+1}).$$

In particular, if we set $\boldsymbol{B}_0 \equiv 0$, $\boldsymbol{\Sigma}_v \equiv 0$, then $\boldsymbol{B}_t \equiv 0$, all t, and this generates a *linear model* with purely (random) time-dependent coefficients. In this case the matrix $\boldsymbol{G}^*_{t-1}$ may be taken to be independent of $\Delta\boldsymbol{x}_t$, and (5.2.37) is now virtually identical with the Harrison–Stevens model. However, with general $\boldsymbol{\Sigma}_v$, (5.2.36) allows us to include both time and state dependence into the model parameters, and thus covers both linear and non-linear models with time-dependent parameters.

5.3 NUMERICAL EXAMPLES OF STATE-DEPENDENT MODEL FITTING

We now apply the algorithm described in Section 5.2 to simulated time series data generated by a variety of non-linear structures. In all cases studied the estimated parameters were smoothed using the non-parametric function fitting technique of Priestley and Chao (1972) using a Gaussian smoothing kernel. This method is usually used to smooth equally spaced ordinates into some functional form; however, in the SDM case, the ordinates are no longer equally spaced since estimated parameter values are obtained at values of successive *states*, not successive time points. This means that if there are one or two outlying states in any set of data excessive importance may be attached to the parameter values at these states by the Gaussian smoothing kernel, and spurious end effects can result. Difficulties can also arise from the "transient effect" at the start of the recursion, where some anomalous parameter values may be obtained before the recursion settles down to a stable state. For these reasons, the first few parameter values in the recursion, and the extreme edges of the parameter curves should be ignored in the following analysis. (Ignoring the first 5% of parameter values and 5% of the values at each end of the parameter curve seems to be a reasonable practical procedure.) Erratic behaviour at the ends of the range can arise also due to the presence of outliers in the original time series data. When an outlier occurs the algorithm tries to accommodate this observation by making large changes in the model parameters—which, in turn, leads to "outliers" in the graph of the estimated parameters. It is important to recall that the SDM algorithm will not necessarily reproduce the "natural" parameter forms for the reasons explained in section (5.2). For the general non-linear model

$$X_t = h(X_{t-1}, \ldots, X_{t-k}, e_{t-1}, \ldots, e_{t-l}) + e_t, \tag{5.3.1}$$

the coefficients $\phi_1, \ldots, \phi_k, \psi_1, \ldots, \psi_l$, will effectively follow the form of the first partial derivatives of h with respect to $X_{t-1}, \ldots, X_{t-k}, e_{t-1}, \ldots, e_{t-l}$, i.e. we may write

$$\phi_u \sim -\frac{\partial h}{\partial X_{t-u}}, \qquad \psi_u \sim \frac{\partial h}{\partial e_{t-u}}.$$

We recall that for the simple threshold AR model (4.2.7), this does, in fact, reproduce the "natural" form of the parameters.

However, for the exponential AR model ϕ_1 does not correspond to its "natural" form, and for the bilinear model the $\{\psi_u\}$ follow their "natural" forms but the $\{\phi_u\}$ contain an additional term, linear in $\{e_{t-u}\}$. (This result highlights the fact that the notion of a "natural" functional form for the model parameters is rather misleading; in the case of, for example, the bilinear model (4.1.1), we may obtain two apparently different representations for the $\{\phi_u\}$ and $\{\psi_u\}$ simply by transferring the bilinear term

$$\left\{\sum_{i=1}^{m} \sum_{j=1}^{k} b_{ij} e_{t-j} X_{t-i}\right\}$$

from the RHS to the LHS of (4.1.1). However, the form of the parameters as determined by a local linearization is free from this ambiguity.)

The form of the parameter μ will, in general, bear little relationship to the "natural" form indicated by the model. It is readily seen that for the general non-linear model (5.3.1)

$$\mu \sim h(\boldsymbol{x}_{t-1}) - \sum_{u=1}^{k} X_{t-u}\left\{\frac{\partial h}{\partial X_{t-u}}\right\}_{\boldsymbol{x}_{t-1}} - \sum_{u=1}^{l} e_{t-u}\left\{\frac{\partial h}{\partial e_{t-u}}\right\}_{\boldsymbol{x}_{t-1}}.$$

Hence, for the first-order exponential AR model (cf. (4.3.11) with $k=1$), for which $h(X_{t-1}) = (\phi_1 + \pi_1 e^{-\gamma X_{t-1}^2})X_{t-1}$, we find

$$\mu \sim 2\gamma X_{t-1}^3 e^{-\gamma X_{t-1}^2},$$

whereas the form in which the model is written down suggests that the natural form for μ should be zero (for all $\boldsymbol{x}_{t-1}$). For this reason graphs of $\hat{\mu}$ for the simulated models are not given.

The following examples are taken from Haggan *et al.* (1984). For each of the models analysed, the range of smoothing in the Priestley-Chao procedure was determined by choosing a Gaussian kernel whose "standard deviation" was set equal to 10% of the total range of the abscissa variable.

1. *Bilinear model*

The algorithm was first applied to data simulated using a bilinear scheme. The model chosen was

$$X_t + 0.4X_{t-1} = 0.5 + 0.4X_{t-1}e_{t-1} + e_t.$$

In all the models, $\{e_t\}$ is a sequence of independent $N(0, 1)$ variables, and 500 observations were simulated on each series. This type of model was chosen, for simplicity, because the functional forms of most of the parameters are functions of one variable only. The recursion procedure described in Section 5.2 is effected by fitting a linear ARMA model to the first 50 observations and starting the recursion at time origin $t=25$. The "smoothing factor" α is taken to be 10^{-4}. If X_t is written in the form

$$X_t = h(X_{t-1}, e_{t-1}) + e_t,$$

and h is differentiated w.r.t. X_{t-1}, e_{t-1}, it may be seen that the functional forms of ψ_1 and ϕ_1 are given by

$$\psi_1 = 0.4X_{t-1}$$

$$\phi_1 = 0.4 - 0.4e_{t-1}.$$

The initial model fitting gave preliminary estimates,

$$\hat{\hat{\mu}} = 0.8800, \qquad \hat{\hat{\phi}}_1 = 0.1777, \qquad \hat{\hat{\psi}}_1 = -0.1427, \qquad \hat{\hat{\sigma}}_e^2 = 1.7663,$$

$$\hat{\hat{\boldsymbol{R}}} = \begin{bmatrix} 0.1197 & 0.1147 & 0.1134 \\ 0.1147 & 0.1157 & 0.1540 \\ 0.1134 & 0.1540 & 0.1711 \end{bmatrix}$$

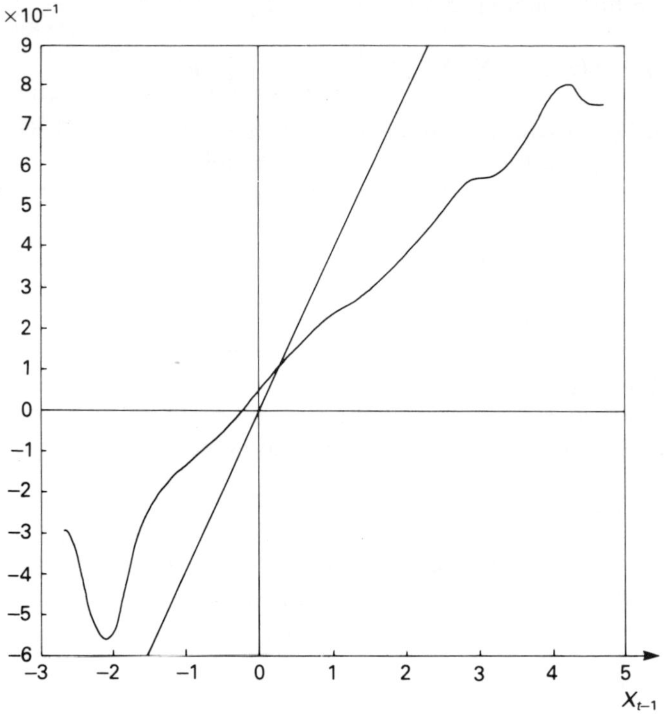

Fig. 5.1. Graph of the parameter $\hat{\psi}_1$ for bilinear model.

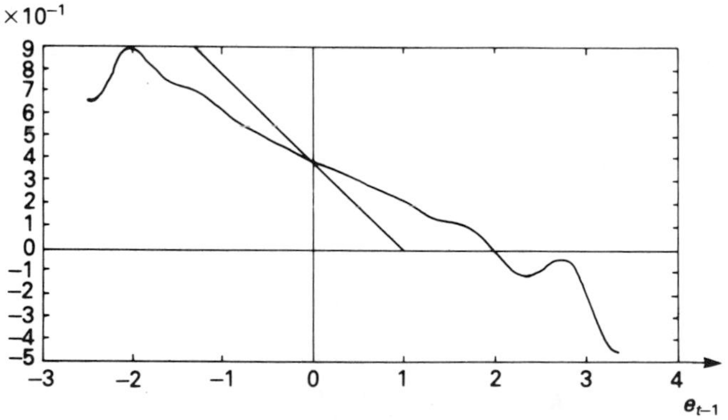

Fig. 5.2. Graph of the parameter $\hat{\phi}_1$ against e_{t-1} for bilinear model.

Figures 5.1 and 5.2 show the resulting graphs of the estimated parameters $\hat{\psi}_1$ and $\hat{\phi}_1$, respectively. The straight lines in these figures represent the true functional forms of ψ_1 and ϕ_1 for comparison.

As may be seen, the slopes of the graphs of the estimated parameters are very close to the slopes of the true parameters, and the intercept of $\hat{\phi}_1$ is also very close to the true value. The intercept of the graph of $\hat{\psi}_1$ is not so close to the true value, but both graphs clearly indicate linear trends in the parameters, and hence provide strong evidence that the series is generated by a bilinear model.

It should be emphasized that the *SDM algorithm operates purely on the data*, and has no *a priori* knowledge of the underlying model. Thus, based on the data only, the SDM algorithm has here pointed us very clearly in the direction of bilinear models.

The reason why the intercept of $\hat{\phi}_1$ is so close to the true value may be due to the fact that the intercept represents the linear part of the model, being simply the autoregressive coefficient of X_{t-1}. In both graphs, as in many other cases studied, the slope of the estimated parameters was slightly less than the slope of the true parameters. This implies that the linear part of this model is estimated very accurately by this algorithm, as one would expect, but the bilinear part is estimated slightly less accurately, again as one would expect, because of the inherent difficulty of bilinear estimation.

2. Exponential AR model

The procedure becomes slightly easier in this case, in that only an AR model fitting procedure is needed for the estimation of initial values. This means that, in general, it is relatively easy to obtain sufficiently accurate initial values using only a small number of observations at the start of the data set. Five hundred observations were simulated on the model

$$X_t+(-0.9-0.1e^{-X_{t-1}^2})X_{t-1}+(0.2+0.1e^{-X_{t-1}^2})X_{t-2}=e_t$$

where $\{e_t\}$ is again a sequence of independent $N(0, 1)$ variables. To initialize the recursive estimation procedure a linear AR model was fitted to the first 20 observations of each series, providing estimate $\hat{\mu}$, $\hat{\phi}_1$, $\hat{\phi}_2$ and $\hat{\sigma}_e^2$ from which to start the recursion at $t=10$. The "smoothing factor" used as 10^{-5}. In this case the functional forms of ϕ_1 and ϕ_2 are given by

$$\phi_1=-\frac{\partial h}{\partial X_{t-1}}=-0.9-0.1e^{-X_{t-1}^2}+0.2X_{t-1}^2e^{-X_{t-1}^2}-0.2X_{t-1}X_{t-2}e^{-X_{t-1}^2}$$

$$\phi_2=-\frac{\partial h}{\partial X_{t-2}}=0.2+0.1e^{-X_{t-1}^2}.$$

The rough functional forms of ϕ_1 and ϕ_2 are shown graphically in Figs. 5.3 and 5.4, respectively.

Preliminary estimates were

$$\hat{\hat{\mu}} = 0.2879, \qquad \hat{\hat{\phi}}_1 = -0.7001, \qquad \hat{\hat{\phi}}_2 = -0.0330, \qquad \hat{\hat{\sigma}}_e^2 = 1.0087,$$

$$\hat{\hat{\boldsymbol{R}}} = \begin{bmatrix} 0.0622 & 0.0160 & -0.0066 \\ 0.0160 & 0.0559 & -0.0418 \\ -0.0066 & -0.0418 & 0.0491 \end{bmatrix}.$$

The graphs of the estimated values $\hat{\phi}_1$ and $\hat{\phi}_2$ are shown in Figs 5.5 and 5.6, respectively. As can be seen, the values of the estimated parameters are approximately correct and the shapes of the graphs are of a similar form to those of the true graphs. $\hat{\phi}_1$ and $\hat{\phi}_2$ are clearly conspicuously different from what might be expected from a bilinear model simulation. Similar results were obtained from several other simulations on this model. A particular feature of the estimated parameters is their approximately symmetric appearance.

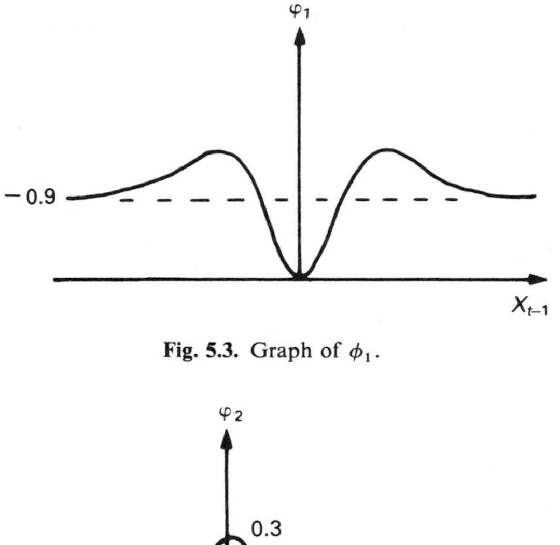

Fig. 5.3. Graph of ϕ_1.

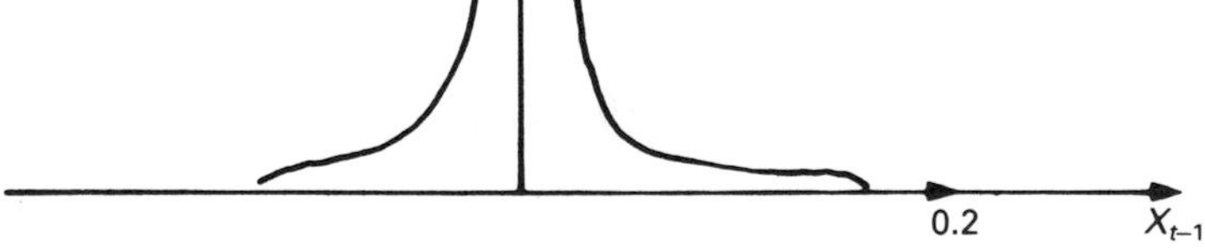

Fig. 5.4. Graph of ϕ_2.

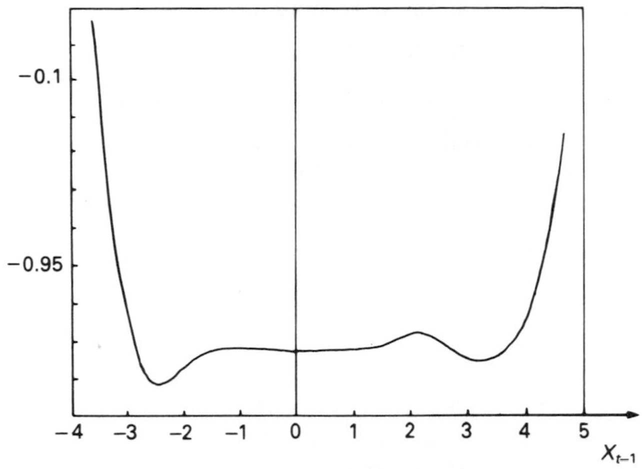

Fig. 5.5. $\hat{\phi}_1$ for exponential model.

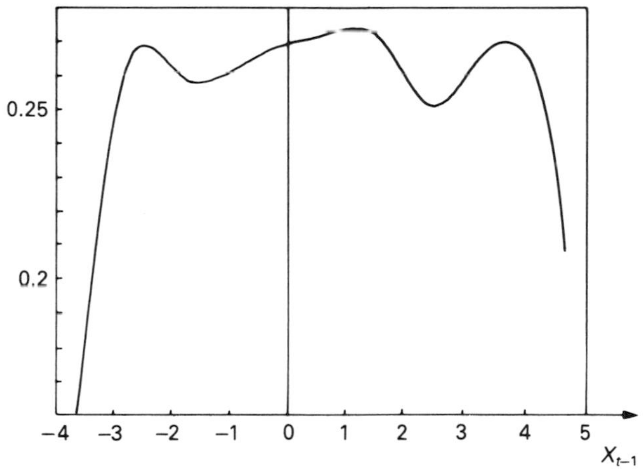

Fig. 5.6. $\hat{\phi}_2$ for exponential model.

3. "Non-linear" threshold AR model

"Non-linear" threshold AR models were introduced by Ozaki (1981) as a modified form of Tong's threshold AR models. These "non-linear" threshold models typically take the form,

$$X_t + (\phi_1 + \pi_1|X_{t-1}|)X_{t-1} + \cdots + (\phi_k + \pi_k|X_{t-1}|)X_{t-k} = e_t, \qquad \text{if } |X_{t-1}| \leq c,$$

$$X_t + (\phi_1 + \pi_1 c)X_{t-1} + \cdots + (\phi_k + \pi_k c)X_{t-k} = e_t, \qquad \text{if } |X_{t-1}| > c.$$

In this study, 500 observations were simulated on the first-order model

$$X_t + (0.1 + 0.4|X_{t-1}|)X_{t-1} = e_t, \qquad \text{if } |X_{t-1}| \leq 1$$
$$X_t + 0.5X_{t-1} = e_t, \qquad \text{if } |X_{t-1}| > 1,$$

with $\{e_t\}$ a sequence of independent $N(0, 1)$ variables.

The functional form of ϕ_1 is given by

$$\phi_1 = -\partial h/\partial X_{t-1} = \begin{cases} 0.1 + 0.8X_{t-1} & \text{if } 0 < X_{t-1} < 1 \\ 0.1 - 0.8X_{t-1} & \text{if } -1 < X_{t-1} < 0 \\ 0.5 & \text{if } X_{t-1} > 1. \end{cases}$$

Graphically, the functional form of ϕ_1 is shown in Fig. 5.7. Note that ϕ_1 is not defined at $X_{t-1} = 1$ and is not differentiable at $X_{t-1} = 0$ and so, for this case, ϕ_1 is not an analytic function. However, it is clearly feasible to approximate the non-linear threshold model by an SDM, so that the estimation algorithm will still be appropriate in this case. The choice of initial estimates for the model fitting procedure was provided by fitting an AR model to the first 20 observations and starting the recursion at $t = 10$. The "smoothing factor" used in the recursion was 10^{-2}.

Preliminary estimates were

$$\hat{\hat{\mu}} = 0.2623, \qquad \hat{\hat{\phi}}_1 = 0.6317, \qquad \hat{\hat{\sigma}}_e^2 = 1.0188$$

$$\hat{\hat{\boldsymbol{R}}} = \begin{bmatrix} 0.0541 & 0.0038 \\ 0.0038 & 0.0297 \end{bmatrix}.$$

The graph of the estimated parameter $\hat{\phi}_1$ is given in Fig. 5.8. This shows that the shape of the parameter $\hat{\phi}_1$ is approximately correct, although the

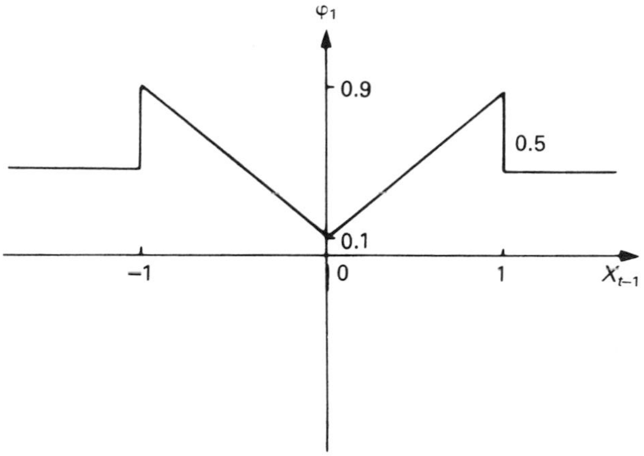

Fig. 5.7. Graph of ϕ_1 for non-linear threshold model.

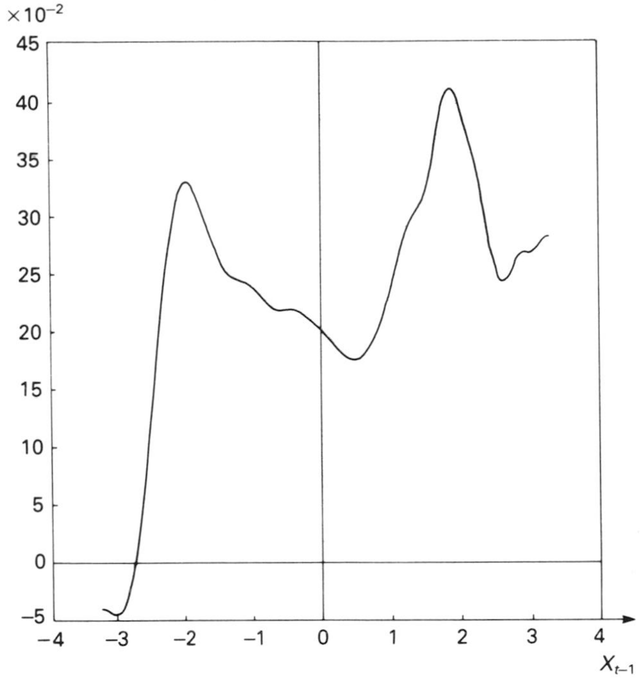

Fig. 5.8. Graph of the parameter $\hat{\phi}_1$ for non-linear threshold model.

actual values taken by the parameter are not very accurate. This feature was also noticeable in other simulations of this type of model, and may probably be accounted for by the fact that the SDM is only an approximation to the non-linear threshold model. In particular, it is very difficult for the algorithm to accommodate the sudden jumps in the parameters at $|X_{t-1}|=1$, and this most probably leads to the depression of the parameter values found in that neighbourhood. However, the results on simulations of this model are encouraging, since the estimated parameters have approximately the correct form and are clearly distinguishable from the estimated parameters of other types of model.

4. *Threshold AR model*

In this case, 500 observations were simulated from the model

$$X_t + 0.5X_{t-1} = 2 + e_t, \qquad \text{if } X_{t-1} \leq 1$$

$$X_t - 0.4X_{t-1} = 0.5 + e_t, \qquad \text{if } X_{t-1} > 1$$

where again $\{e_t\}$ is a sequence of independent $N(0, 1)$ variables. As for the simulations 2 and 3, AR model fitting to the first 20 observations was used to provide initial estimates for the recursion which was started at $t = 10$. In this case, the smoothing factor used was 10^{-5}.

Here we have

$$\phi_1 = \begin{cases} 0.5 & \text{if } X_{t-1} \leq 1, \\ -0.4 & \text{if } X_{t-1} > 1, \end{cases}$$

so that ϕ_1 is not an analytic function of X_{t-1}, since it is undefined at $X_{t-1} = 1$. Thus, the SDM provides only an approximation to this model.

Preliminary estimates were

$$\hat{\hat{\mu}} = 1.7649, \qquad \hat{\hat{\phi}}_1 = 0.0110, \qquad \hat{\hat{\sigma}}_e^2 = 0.9926,$$

$$\hat{\hat{\boldsymbol{R}}} = \begin{bmatrix} 0.1803 & 0.0760 \\ 0.0760 & 0.0452 \end{bmatrix}.$$

However, in this case the SDM algorithm does not give as clear an indication of the model as in the previous examples, as may be seen from the graph of the estimated parameter $\hat{\phi}_1$ shown in Fig. 5.9.

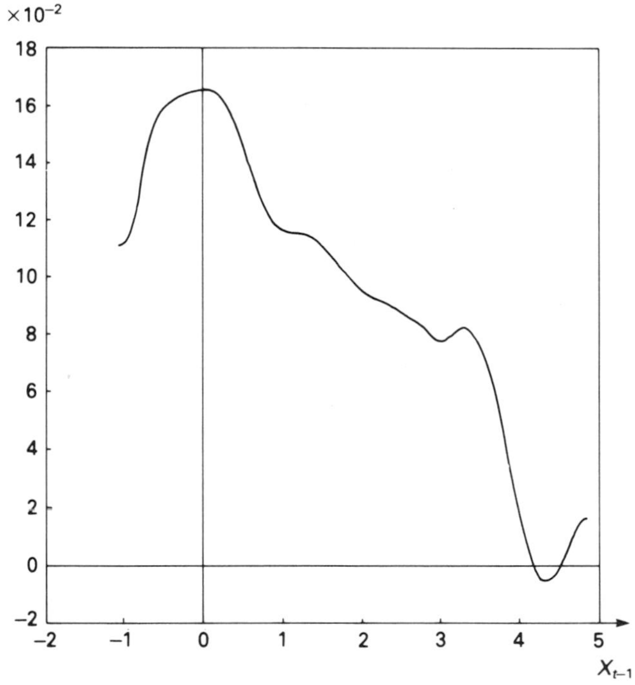

Fig. 5.9. Graph of the parameter $\hat{\phi}_1$ for the threshold AR model.

The estimated parameter $\hat{\phi}_1$ in this case takes values midway between the values 0.5 and −0.4 and there is no clear sign of a jump in $\hat{\phi}_1$ at $X_{t-1}=1$. It appears that the SDM algorithm, in its original form, is not efficient in detecting the sudden change in coefficients inherent in the "linear" threshold model, although it is capable of detecting the slower change of coefficients as specified by the "non-linear" threshold model. However, if we modify the algorithm as described below we obtain a much clearer indication of the threshold effect.

It is readily seen that for the above threshold model the series describing the gradient parameter $\gamma_1^{(t)}$ consists entirely of negative (and zero) values, and consequently an obvious modification is to replace the random walk model by a non-zero mean AR(1) model of the form,

$$\gamma_1^{(t)}=\nu+\lambda\gamma_1^{(t-1)}+V_t, \qquad (|\lambda|<1). \tag{5.3.2}$$

(This modification is easily accommodated within the general SDM algorithm by adding the additional element unity to the "state vector" $\boldsymbol{\theta}_t$, and similarly enlarging the matrix $\boldsymbol{F}_t^*$. The full details are given in Haggan *et al.* (1984), together with a discussion of the general behaviour of the gradient parameter estimates.) To illustrate the improved efficiency arising from the modified gradient parameter model, the data for the threshold AR model described in example 4 were reanalysed using an AR model of the form (5.3.2) for the gradient parameter $\gamma_1^{(t)}$. The resulting form of $\hat{\phi}_1$ is shown in Fig. 5.10. It will be seen that the new graph of $\hat{\phi}_1$ gives a much more accurate indication of the true form of this parameter, and the "threshold" effect is now fairly obvious.

Sampling properties

To give some indication of the sampling variability of the estimated parameter functions, several independent realizations (each of length 500) were generated from particular models, and the same SDM procedure applied to each realization.

Figures 5.11 and 5.12 show the resulting graphs of the estimated parameters $\hat{\psi}_1$ (as a function of X_{t-1}) and $\hat{\phi}_1$ (as a function of e_{t-1}) for three different realizations of the bilinear model used in example 1, namely

$$X_t+0.4X_{t-1}=0.5+0.4X_{t-1}e_{t-1}+e_t.$$

The smoothing factor used was 10^{-4} in each case. (The straight lines in these figures represent the true functional forms of ψ_1 and ϕ_1.) As may be seen, both the slopes and the intercepts of the graphs of the estimated parameters are very close to those of the true parameters. Moreover, the variability over the three realizations is relatively small, and does not in

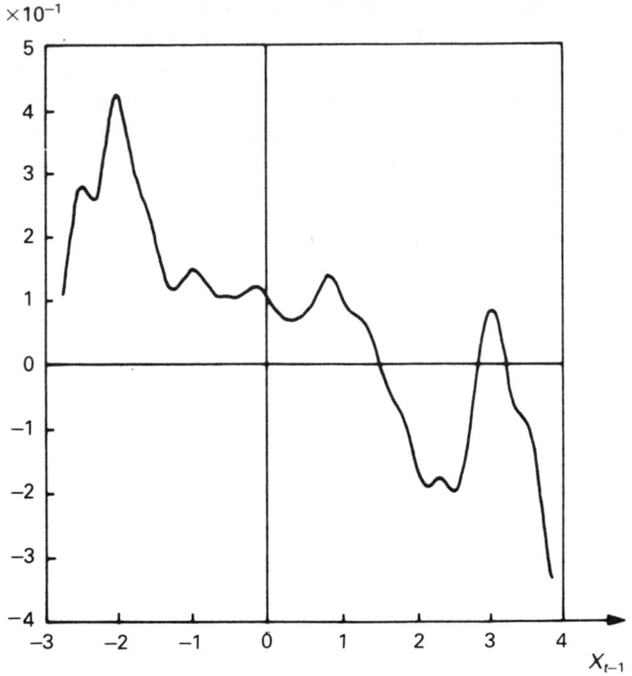

Fig. 5.10. Revised form of $\hat{\phi}_1$ for threshold model.

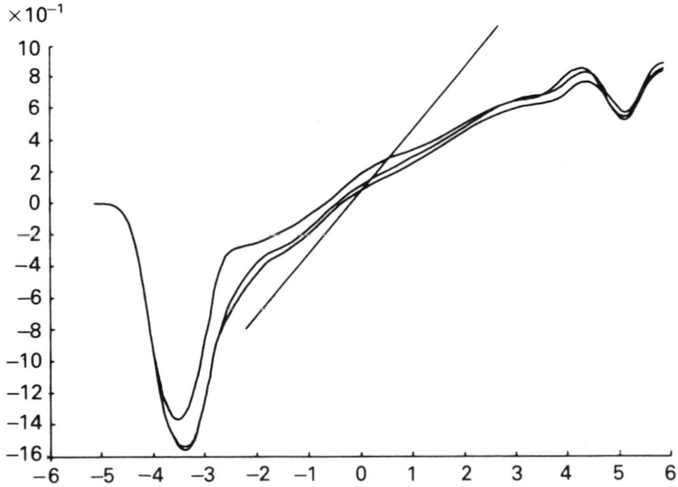

Fig. 5.11. Graphs of $\hat{\psi}_1$ against X_{t-1} for bilinear model.

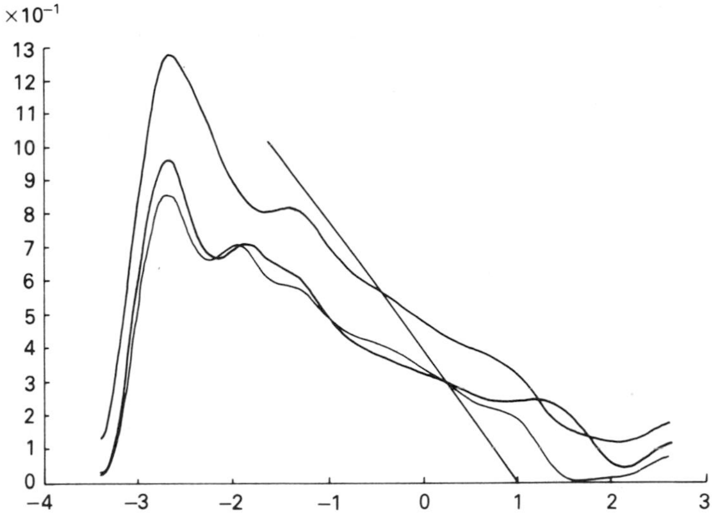

Fig. 5.12. Graphs of $\hat{\phi}_1$ against e_{t-1} for bilinear model.

any way obscure the clear indication of linear trends in the parameters—providing strong evidence that the series is generated by a bilinear model. As a further example, Fig. 5.13 shows the graphs of $\hat{\phi}_1$ (as a function of

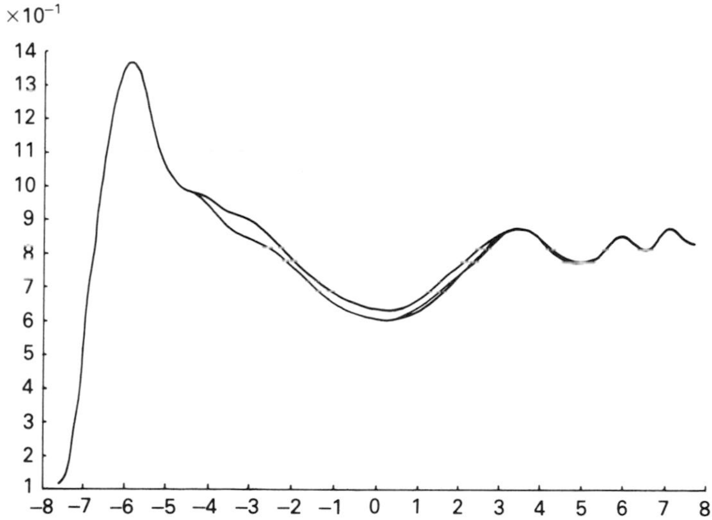

Fig. 5.13. Graphs of $\hat{\phi}_1$ against X_{t-1} for exponential model (5.3.3).

X_{t-1}) for three different realizations of the exponential model

$$X_t + (0.9 - 0.2e^{-X_{t-1}^2})X_{t-1} = e_t \tag{5.3.3}$$

using an AR model of the form (5.3.2) for the gradient parameters. The smoothing factor was 10^{-3}. Again, it can be seen that the shapes of the estimated parameter function for three different realizations are all very similar, the sampling functions being almost negligible.

Except for large X_{t-1}, the values of $\hat{\phi}_1$ are approximately correct and the shapes of the graphs are of a similar form to the true graph (Fig. 5.14) given by,

$$\phi_1 = \frac{-\partial h}{\partial X_{t-1}} = 0.9 - 0.2\, e^{-X_{t-1}^2} + 0.4X_{t-1}^2\, e^{-X_{t-1}^2}.$$

Figure 5.15 also shows the graph of variance $\hat{\phi}_1$, against t, derived from the Kalman algorithm described in Section 5.2. The variance of $\hat{\phi}_1$ is small throughout the range, and after approximately 10 time units it settles down to a steady value of approximately 0.002.

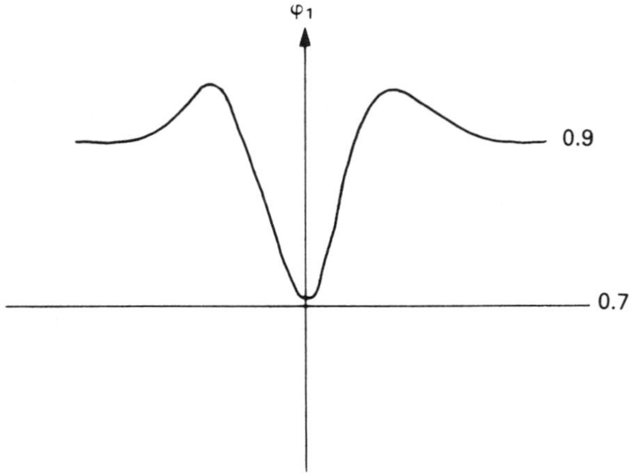

Fig. 5.14. Graph of ϕ_1 against X_{t-1} for exponential model (5.3.3).

Analysis of Canadian lynx series and sunspot series

1. *Candian lynx series*

As noted in Section 4.2, the behaviour of the Canadian lynx series is indicative of non-linearity since (i) the time spent in rising from "trough"

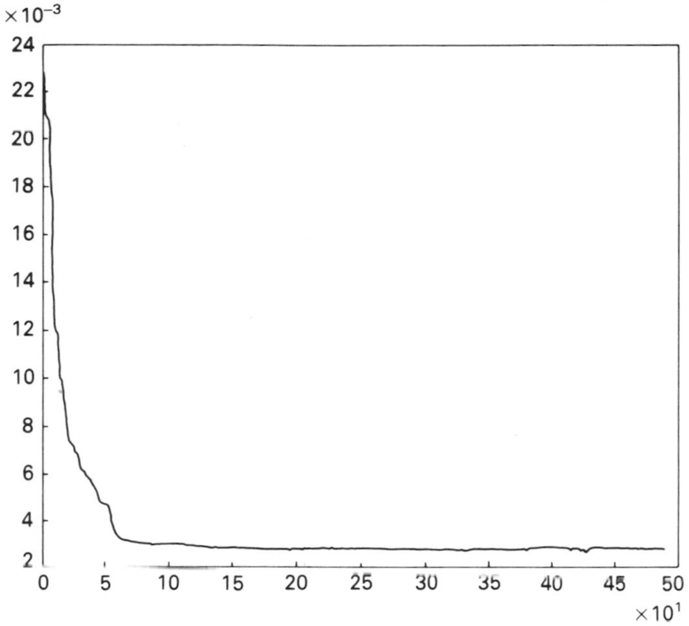

Fig. 5.15. Graph of variance $\hat{\phi}_1$ for exponential model (5.3.3).

to "peak" appears in general to be longer than the time taken to fall from "peak" to "trough", and (ii) in the study of animal populations it has been asserted (Bulmer, 1974; Tong, 1983) that many such populations display "limit cycle" behaviour, which is a strictly non-linear phenomenon. In the analysis carried out by Tong and Lim (1980) and Haggan and Ozaki (1981), it was shown that both a (linear) threshold AR and an exponential AR model provided a good fit to the Canadian lynx data. Both of these models also indicated the presence of a limit cycle of period of about 9–9.5 years. However, it may be surmised, e.g., from examination of the characteristic roots of the fitted exponential AR model (Haggan and Ozaki, 1981), that the departure from non-linearity is not very large. It may be noted further that if one were to discretize a sine curve of period 9–9.5 years at yearly intervals for 114 years, then it may appear that the rise and fall times for the discretized data are consistently different. The mean difference between "rise" and "fall" times in the lynx series is not much greater than might be expected for a discretized sine curve. Hence, this may not be a sufficient reason to indicate the fitting of a non-linear model. However, the possibility of determining some form of limit cycle behaviour is appealing, in this context, and this leads us to consider some form of non-linear model.

An SDM of AR(2) form was fitted, for simplicity, to the data (logarithmically transformed), and the resulting parameters ϕ_1 and ϕ_2 were estimated in the manner previously described, with smoothing factor equal to 10^{-5}.

Preliminary estimates were

$$\hat{\hat{\mu}} = 1.22373, \qquad \hat{\hat{\phi}}_1 = -1.3105, \qquad \hat{\hat{\phi}}_2 = 0.7218, \qquad \hat{\hat{\sigma}}_e^2 = 0.0661,$$

$$\hat{\hat{\boldsymbol{R}}} = \begin{bmatrix} 0.1396 & 0.0270 & 0.0177 \\ 0.0270 & 0.0359 & -0.0274 \\ 0.0177 & -0.0274 & 0.0338 \end{bmatrix}.$$

The graphs of $\hat{\phi}_1$ and $\hat{\phi}_2$ are shown in Figs 5.16 and 5.17.

As may be seen, the parameters show a degree of oscillation about some fairly constant value, the values being approximately the same as those obtained by Moran (1953) in his AR(2) model fitting. This is in agreement with the previous observation that the departure from non-linearity of the lynx series is not very great. The general shape of the parameters $\hat{\phi}_1$ and $\hat{\phi}_2$ would seem to indicate that if a non-linear model is to be fitted, an

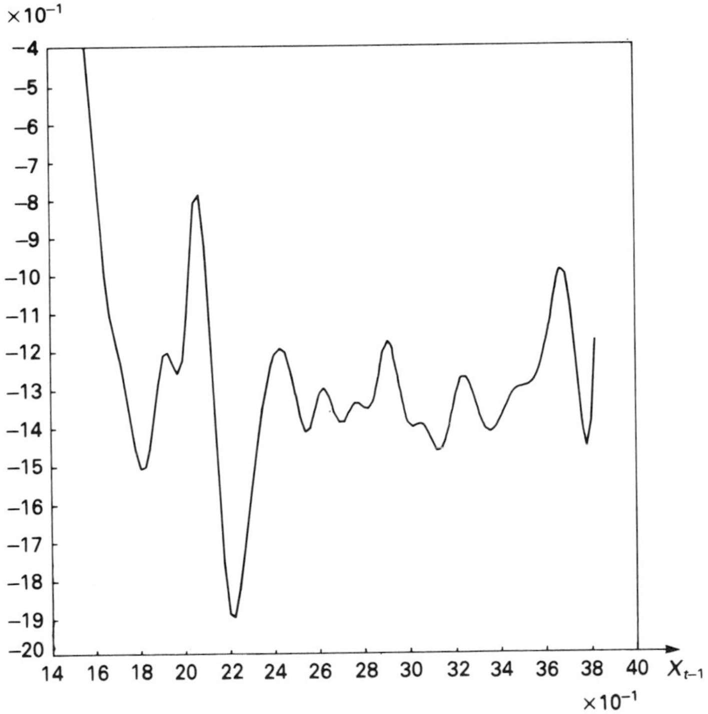

Fig. 5.16. Graph of the parameter $\hat{\phi}_1$ for lynx data.

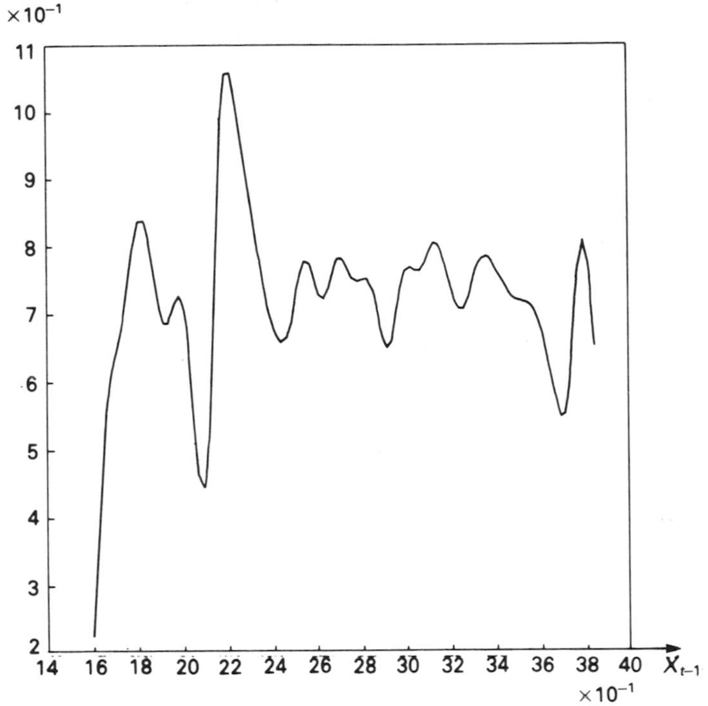

Fig. 5.17. Graph of the parameter $\hat{\phi}_2$ for lynx data.

exponential AR model, rather than a non-linear threshold AR model or bilinear model, might be appropriate. It should, however, be noted that Tong and Lim (1980) have fitted this data very well using a (linear) threshold AR model. As discussed previously, the SDM algorithm, in its basic form, is not particularly efficient in detecting threshold AR models, and so the fact that the threshold AR model has not been indicated in this case is certainly not conclusive.

One noticeable feature in the graph of both $\hat{\phi}_1$ and $\hat{\phi}_2$ is the sharp "dip" which occurs in the neighbourhood of $X_{t-1} = 2.2$. Since the dip is clearly present in both graphs, it would seem to correspond to some real feature of the data (rather than an artefact of the algorithm), but the value 2.2 does not appear to correspond to any plausible "threshold" value for this series.

A futher feature of the graphs is the very high negative "correlation" between the fluctuations in $\hat{\phi}_1$ and $\hat{\phi}_2$. This is a puzzling result, and it persists throughout the application of varying levels of the smoothing factor. One possible explanation is that the sum of the two parameters, $(\phi_1 + \phi_2)$, is stable throughout the process, and is thus well estimated by the algorithm. However, we feel that this effect may well be due to the short length of the lynx series (114 observations), leading in turn to only a small number of points in the $\hat{\phi}_1$, $\hat{\phi}_2$, graphs which are very unevenly spread over the respective ranges. Smoothing a small number of unevenly spaced points can certainly lead to fluctuations of the above type, and very similar negative "correlation" between the parameters of a second-order exponential model has been observed when analysing a short stretch of 114 observations.

2. *The sunspot series*

As mentioned in Section 4.1, this series consists of 246 observations describing the numbers of sunspots observed annually between 1700 and 1945. The data is highly asymmetric and has not proved amenable to linear analysis even when logarithmically transformed. One of the simplest "pseudo-periodic" linear time series models is the AR(2) model, and to

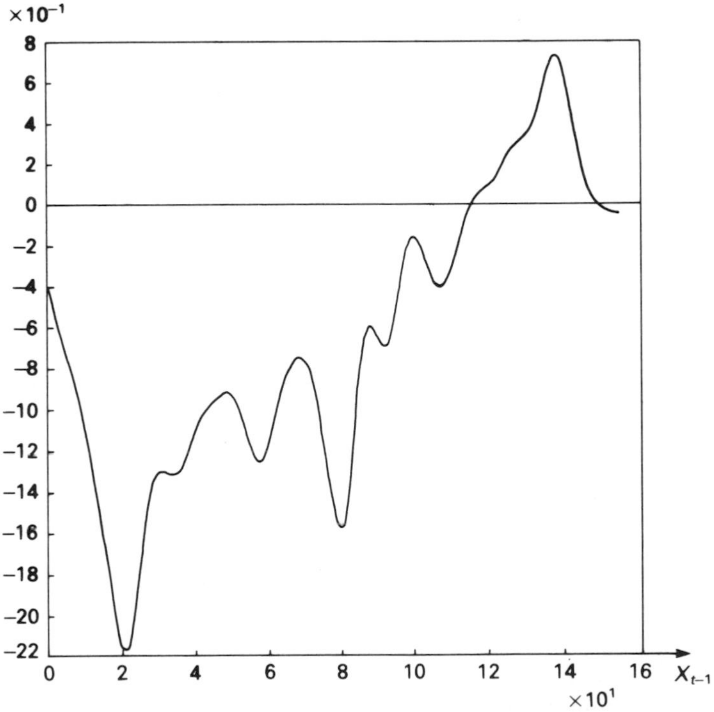

Fig. 5.18. Graph of the parameter $\hat{\phi}_1$ for sunspot data.

analyse the non-linear structure of the model, an SDM of AR(2) form was fitted to the data in the usual way.

Preliminary estimates were

$$\hat{\mu} = 8.1468, \qquad \hat{\hat{\phi}}_1 = -1.2483, \qquad \hat{\hat{\phi}}_2 = 0.6069, \qquad \hat{\hat{\sigma}}_e^2 = 114.436,$$

$$\hat{\hat{\boldsymbol{R}}} = \begin{bmatrix} 15.106 & 0.2421 & 0.1484 \\ 0.2421 & 0.0444 & -0.0392 \\ 0.1484 & -0.0392 & 0.0523 \end{bmatrix}.$$

Using a smoothing factor of 10^{-2}, the estimates $\hat{\phi}_1$ and $\hat{\phi}_2$ are shown in Figs 5.18 and 5.19. These estimates both show a fairly strong linear trend indicating that a bilinear model may be most appropriate in this case. This concurs with the results of Gabr and Subba Rao (1981), who have successfully fitted bilinear models to this series.

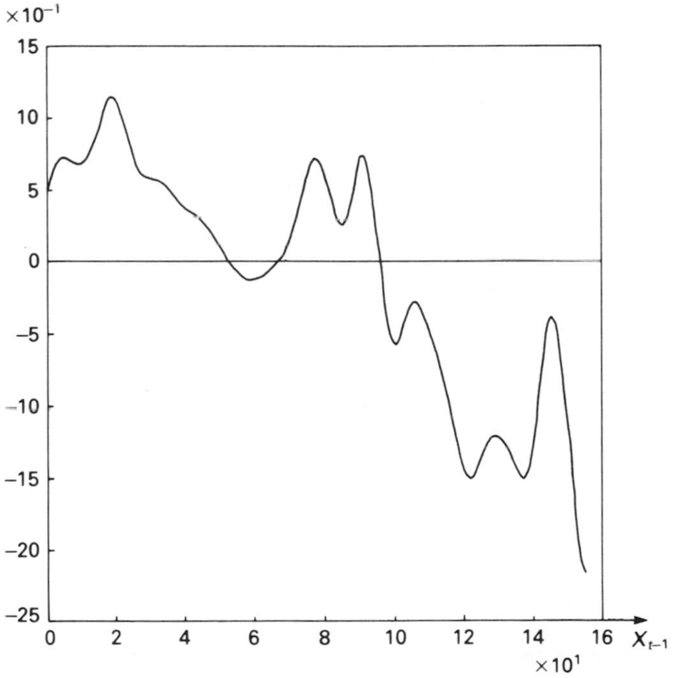

Fig. 5.19. Graph of the parameter ϕ_2 for sunspot data.

5.4 SDM SCHEMES FOR NON-LINEAR SYSTEMS

It is well known that a conventional linear time series model may be interpreted as a "dynamical systems" model in which the observed series is regarded as the output of a linear system "driven" by an (unobservable) white noise input, $\{e_t\}$. Thus, if we are dealing with a physical system in which the input $\{U_t\}$ and the output $\{X_t\}$ are both observable, we may adapt a linear time series model to this situation by replacing the white noise process $\{e_t\}$ by the observable input process $\{U_r\}$ (see, for example, *Spec. Anal.*, Chapter 10). The same approach may be applied to the non-linear SDM schemes. Consider the single input/single output systems shown schematically in Fig. 5.20.

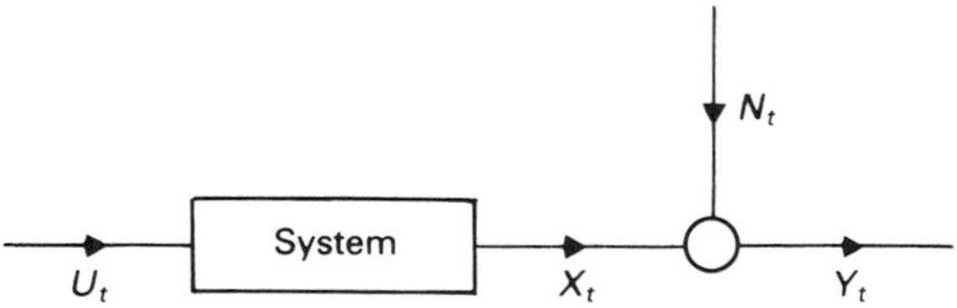

Fig. 5.20. Block diagram showing single input/single output systems.

In Fig. 5.20, U_t denotes the input, X_t the "uncorrupted" output, N_t is an additive noise disturbance, and Y_t is the observed output. We assume that the system is operating in "open loop" form, so that it is reasonable to assume that $\{N_t\}$ and $\{U_t\}$ are uncorrelated processes. The "systems" model corresponding to (5.1.1) takes the form,

$$X_t = h(X_{t-1}, X_{t-2}, \ldots, X_{t-k}, U_{t-1}, U_{t-2}, \ldots, U_{t-l})$$
$$+ g(X_{t-1}, X_{t-2}, \ldots, X_{t-k}, U_{t-1}, U_{t-2}, \ldots, U_{t-l})U_t \quad (5.4.1)$$

together with the observation equation,

$$Y_t = X_t + N_t. \quad (5.4.2)$$

Equation (5.4.1) may now be rewritten in SDM form as

$$X_t + \phi_1(\boldsymbol{x}_{t-1})X_{t-1} + \cdots + \phi_k(\boldsymbol{x}_{t-1})X_{t-k}$$
$$= \mu(\boldsymbol{x}_{t-1}) + \psi_0(\boldsymbol{x}_{t-1})U_t + \psi_1(\boldsymbol{x}_{t-1})U_{t-1} + \cdots + \psi_l(\boldsymbol{x}_{t-1})U_{t-l} \quad (5.4.3)$$

where the state vector at time $(t-1)$ is here given by

$$\boldsymbol{x}_{t-1} = (U_{t-l}, \ldots, U_{t-1}, X_{t-k}, \ldots, X_{t-1})' \quad (5.4.4)$$

(Note that since U_t is an observable process, the coefficient of U_t, ψ_0, is no longer unity but is now an additional parameter of the model.) If we again augment the state vector by including the constant unity as an additional element, i.e. if we now write

$$\boldsymbol{x}_{t-1} = (1, U_{t-l}, \ldots, U_{t-1}, X_{t-k}, \ldots, X_{t-1})', \tag{5.4.5}$$

then we can write (5.4.2), (5.4.3) in the following state-space form,

$$\boldsymbol{x}_{t+1} = \{\boldsymbol{F}(\boldsymbol{x}_t)\}\boldsymbol{x}_t + \{\boldsymbol{G}(\boldsymbol{x}_t)\}U_{t+1} \tag{5.4.6}$$

$$Y_t = \boldsymbol{H}\boldsymbol{x}_t + N_t \tag{5.4.7}$$

where the system matrix $\boldsymbol{F}$ has the same form as (5.2.3), the input matrix $\boldsymbol{G}$ is given by,

$$\boldsymbol{G} = (0, 0, \ldots, 1 \vdots 0, \ldots, \psi_0)',$$

and the observation matrix $\boldsymbol{H}$ by

$$\boldsymbol{H} = (0, 0, \ldots, 0 \vdots 0, \ldots, 1).$$

Assuming, as previously, that the coefficients μ, $\{\phi_u\}$, $\{\psi_u\}$ may be represented as locally linear functions of $\boldsymbol{x}_t$, the "up-dating" equations then take the same form as (5.2.11). (Note, however, that due to the presence of the additional coefficient, ψ_0, there is an additional gradient vector, $\boldsymbol{\beta}_0^{(t)}$.) The gradient parameters $(\boldsymbol{\alpha}^{(t+1)}, \boldsymbol{\beta}_u^{(t+1)}, \boldsymbol{\gamma}_u^{(t+1)})$ can similarly be allowed to wander in the form of random walks, or as stationary AR(1) models. These may be written respectively as

$$\boldsymbol{B}_{t+1} = \boldsymbol{B}_t + \boldsymbol{V}_{t+1},$$

or

$$\boldsymbol{B}_{t+1} = \lambda \boldsymbol{B}_t + \boldsymbol{V}_{t+1}$$

where $\boldsymbol{B}_t$ is the matrix defined in (5.2.10), and $\{\boldsymbol{V}_t\}$ is again a sequence of independent matrix-valued random variables, with $\boldsymbol{V}_t = N(\boldsymbol{0}, \boldsymbol{\Sigma}_v)$. Our estimation procedure then determines, for each t, those values of $\boldsymbol{B}_t$ which minimize the discrepancy between the observed value of Y_{t+1} and its predictor $\hat{Y}_{t+1}$ computed from the model fitted at time t. As in the case of the pure time series models, we implement this recursive fitting by rewriting the state-space representation of the SDM model in terms of a new state vector which represents the current values of all the model parameters.

The full details of this approach (together with a description of the initial model fitting procedure and choice of starting values) are given in Priestley and Heravi (1985). We give below two illustrations of the application of this technique to simulated data generated by (i) a linear system, and (ii) a bilinear system. Further numerical examples are given in Priestley and Heravi (1985).

In each case, 500 observations were generated for the input and output processes, and the input process $\{U_t\}$ was generated from the stationary AR(1) model,

$$U_t + 0.5\,U_{t-1} = e_t,$$

$\{e_t\}$ being a sequence of independent $N(0, 1)$ variables. The noise disturbance $\{N_t\}$ was also generated from a Gaussian white noise process, with zero mean, unit variance.

For each model, a linear ARMA type model was fitted to the first 50 observations, and the recursion started at time $t = 25$. Graphs of the $\{\hat{\phi}_u\}$ and $\{\hat{\psi}_u\}$ were plotted against the corresponding values of X_{t-1} or U_{t-1}, which are then compared with the theoretical functional forms corresponding to each simulated model. It should be noted that, in common with the "time series" case, the SDM algorithm will not necessarily reproduce the "natural" parameter forms, except for the parameter ψ_0.

A. *Linear system simulation*

The particular system chosen was given by,

$$X_t + 0.8X_{t-1} = U_t - 0.6\,U_{t-1}$$

$$Y_t = X_t + N_t.$$

Here, the functional form of ϕ_1 should be equal to 0.8 and ψ_0 and ψ_1 should be equal to 1 and -0.6 respectively.

The preliminary estimates were:

$$\hat{\hat{\mu}} = -0.2242, \qquad \hat{\hat{\phi}}_1 = 0.5558, \qquad \hat{\hat{\psi}}_0 = 1.034, \qquad \hat{\hat{\psi}}_1 = -0.9373,$$

$$\hat{\hat{\sigma}}^2_N = 1.142$$

$$\hat{\hat{\boldsymbol{R}}} = \begin{bmatrix} 0.0256 & -0.0041 & -0.0040 & -0.0028 \\ -0.0041 & 0.0136 & 0.0040 & 0.0027 \\ -0.0040 & 0.0040 & 0.0232 & -0.0107 \\ -0.0028 & 0.0027 & -0.104 & 0.0224 \end{bmatrix}.$$

In this case the smoothing factor α was taken to be 10^{-4} and the resulting graphs of $\hat{\phi}_1$, $\hat{\psi}_0$, $\hat{\psi}_1$ are given in Figs 5.21–5.23. We note that, except for the usual end effects, they are consistent with their theoretical values.

The constancy in the estimated parameters $\hat{\phi}_1$, $\hat{\psi}_0$, $\hat{\psi}_1$ give a clear indication that these series were generated by a linear system.

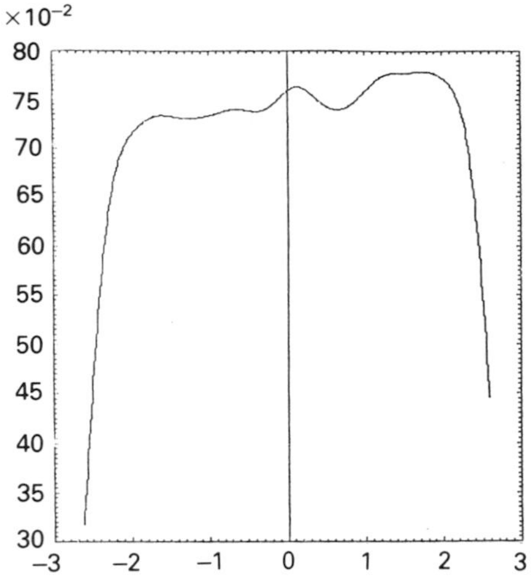

Fig. 5.21. ϕ_1 for model A (linear system).

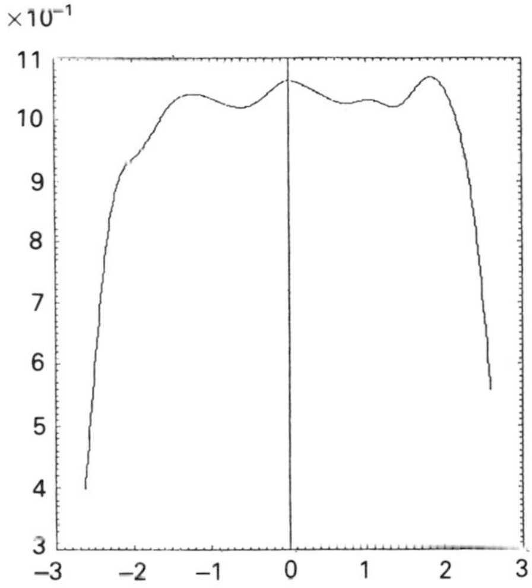

Fig. 5.22. $\hat{\psi}_0$ for model A (linear system).

B. Bilinear system simulation

The system chosen was,

$$X_t - 0.4X_{t-1} = 0.8 + 0.4X_{t-1}U_t$$

$$Y_t = X_t + N_t.$$

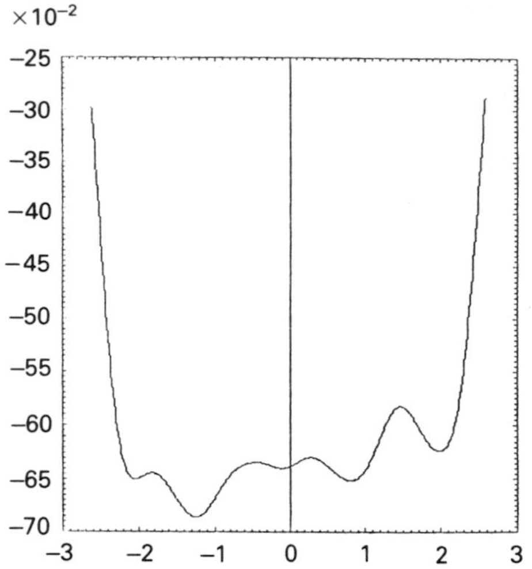

Fig. 5.23. $\hat{\psi}_1$ for model A (linear system).

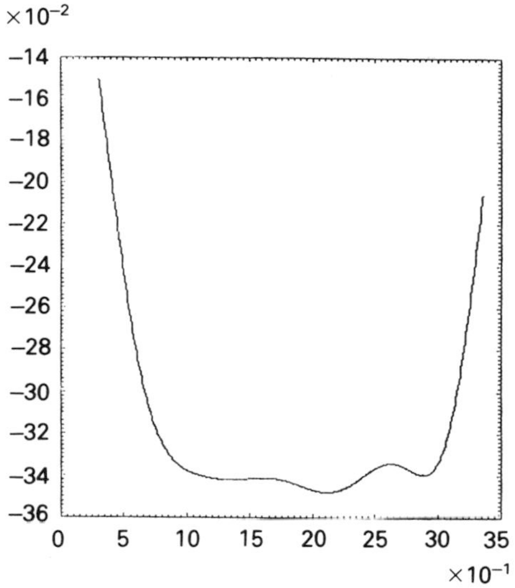

Fig. 5.24. $\hat{\phi}_1$ for model B (bilinear system).

If X_t is written in the form (5.4.1) it can be seen that the functional form of ϕ_1 is given by

$$\phi_1 = -\frac{\partial h}{\partial X_{t-1}} = -0.4$$

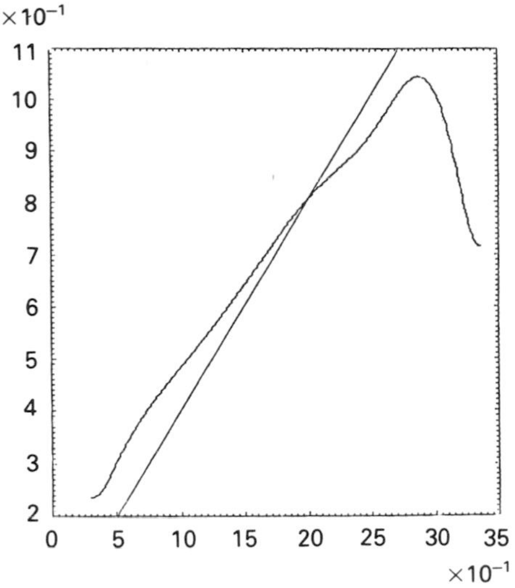

Fig. 5.25. $\hat{\psi}_0$ for model B (bilinear system).

and the functional form of ψ_0 is

$$\psi_0 = g(X_{t-1}) = 0.4X_{t-1}$$

The preliminary estimates were:

$$\hat{\hat{\mu}} = 1.301, \qquad \hat{\hat{\phi}}_1 = -0.3436, \qquad \hat{\hat{\psi}}_0 = 0.8391, \qquad \hat{\hat{\sigma}}^2_N = 1.768$$

$$\hat{\hat{\boldsymbol{R}}} = \begin{bmatrix} 0.1276 & 0.0496 & 0.0200 \\ 0.0496 & 0.0273 & 0.0143 \\ 0.0200 & 0.0143 & 0.0326 \end{bmatrix}.$$

Figures 5.24 and 5.25 show the resulting graphs of ϕ_1 and ψ_0 (as functions of X_{t-1}), respectively, and the straight line in Fig. 5.25 represents the true functional form of ψ_0. Again, apart from the end effects, the estimated values of $\hat{\phi}_1$ are constant, one very close to the value of -0.4 expected for this parameter. The graph of $\hat{\psi}_0$ shows a clear linear trend, with the slope very close to that of the expected functional form. The results thus give a very clear indication of the bilinear form of the underlying system.

Chapter 6

Non-stationary Processes

As remarked in Section 1.2, conventional time series analysis is heavily dependent on the twin assumptions of *linearity* and *stationarity*. In the preceding chapters we discussed the types of models and analyses which are available when the assumption of linearity is dropped. In this and in the following chapter we explore the situation when the assumption of stationarity is relaxed.

It is not difficult to see why the notion of stationarity is such an appealing one: it endows the process with "statistical stability" so that quantities originally defined as ensemble averages (such as, for example, autocovariances and autocorrelations) can be estimated from a single realization by computing the corresponding time domain averages. (Strictly speaking, the property to which we are referring here is "ergodicity" rather than stationarity, but in practical terms the two ideas are very closely related.) However, like virtually all mathematical concepts, stationarity is an idealization, and in practice it can at best be expected to hold only as an approximation. It is clearly of interest, therefore, to examine what types of analyses are available for those cases where the assumption of stationarity becomes unrealistic, and this study forms the basis of the present chapter.

Clearly, if we simply drop the assumption of stationarity and do not replace it with any alternative assumptions then there is very little that we can say about a given series. In order to develop a useful theory we need to replace stationarity by a more general notion which still allows us to carry out meaningful statistical analyses. Here, we may be guided by the approach which we adopted in discussing non-linear "state-dependent" models; in that context we moved away "gradually" from linearity and arrived at a class of models which, although globally non-linear, possessed the property of being "locally linear". In a similar spirit, we may now consider processes which, although globally non-stationary, are in some sense "locally stationary". These ideas lead us to a class of processes termed "semi-stationary", and allow us to develop a form of time-dependent spectral analysis which shares many of the features of the spectral analysis of stationary processes.

6.1 TYPES OF NON-STATIONARY PROCESSES

Non-stationary processes may arise in several ways. One of the simplest situations is that in which the observed process, $\{X_t\}$, takes the form

$$X_t = \mu_t + Z_t \tag{6.1.1}$$

where μ_t is a deterministic function of t, and $\{Z_t\}$ is a zero-mean stationary process. Equation (6.1.1) is the basic "trend plus stationary residual" model, and it simply allows the mean $E[X_t]$ to vary over time in an arbitrary fashion. Typical examples would correspond to taking $\{\mu_t\}$ as (i) a finite-order polynomial in t, describing a steady "growth" or "decay" in the long-term behaviour of the process, or (ii) a finite sum of trigonometric terms, describing cyclical or "seasonal" behaviour (although in the latter case we can manipulate the model into a stationary form). This type of model is fairly easy to deal with: we may either parameterize the function μ_t, estimate its parameters, and then subtract the estimated μ_t and perform all subsequent analyses on the "residuals" Z_t, or alternatively we may remove the mean function μ_t by applying suitable difference operators to X_t—see, for example, *Spec. Anal.*, p. 587).

A more interesting example is that of "explosive" AR or ARMA models. We recall from Section 2.2 that the condition for the ARMA model

$$\alpha(B)X_t = \beta(B)\varepsilon_t \tag{6.1.2}$$

to generate an (asymptotically) stationary process is that all the roots of $\alpha(z)$ lie outside the unit circle. If $\alpha(z)$ has roots on or inside the unit circle then (6.1.2) generates a non-stationary process. (If all the roots are strictly inside it may be possible to construct a stationary solution of (6.1.2) but only with X_t expressed in terms of $\varepsilon_t, \varepsilon_{t+1}, \varepsilon_{t+2}, \ldots$: *Spec. Anal.*, p. 134.) Suppose, for example, that $\alpha(z)$ has a d-fold root at $z=1$, and all other roots lie outside $|z|=1$. We may then write (6.1.2) as

$$(1-B)^d \alpha'(B)X_t = \beta(B)\varepsilon_t \tag{6.1.3}$$

with $\alpha'(B)$ a "stationary" AR operator, which in turn can be rewritten as

$$\alpha'(B)Y_t = \beta(B)\varepsilon_t \tag{6.1.4}$$

with $Y_t = \Delta^d X_t$, $\Delta \equiv (1-B)$ being the usual difference operator. Equation (6.1.3) thus postulates that if we difference the $\{X_t\}$ process d times we obtain a stationary ARMA process $\{Y_t\}$. The non-stationary character of this model may be seen by interpreting (6.1.3) as the result of passing the white noise process $\{\varepsilon_t\}$ through an "unstable" filter whose transfer function, $\beta(e^{-i\omega})/\alpha(e^{-i\omega})$ has a dth-order pole on the unit circle. Models of this form were studied extensively by Box and Jenkins (1970), who described them under the term "ARIMA" (autoregressive integrated moving-average). At first sight (6.1.4) would appear to indicate that X_t is of the form (6.1.1) with μ_t a polynomial of degree $(d-1)$, and Z_t a stationary ARMA process. If X_t is of this form then it is certainly true that its dth difference will be a stationary process, but (6.1.3) allows for a more general structure. The essential point is that if X_t is of the form (6.1.1) then the operator $\beta(B)$ in (6.1.3) will also contain $(1-B)^d$ as a factor: if $\beta(B)$ does not contain this factor then Z_t is itself a non-stationary process (see, for example, *Spec. Anal.*, p. 770 for further discussion).

It should, however, be noted that although the ARIMA scheme generates non-stationary processes, the non-stationarity inherent in these processes is of a rather special form, i.e. it describes "explosive" behaviour. The second-order properties of such processes vary over time, but the complete evolution of the process is governed by *a fixed set of parameters*. Thus, if we fit an ARIMA model to an initial stretch of data the complete future of the process is then determined by the values of the parameters fitted to the initial data. This may seem unduly restrictive if we wish to study processes whose second-order processes vary in an arbitrary fashion over time. The non-stationary character of this latter class is more closely linked to, e.g., ARMA models with freely varying time-dependent parameters rather than fixed parameter models of the explosive type. It is this more general class of non-stationary processes to which we now turn our attention, and in the following sections we describe various approaches to their spectral representation and analysis.

6.2 THE NOTION OF TIME-DEPENDENT SPECTRA

The spectrum (or more precisely, spectral density function), $h(\omega)$, of a stationary process was defined in Section 1.3, and one of its basic properties is that it possesses a physical interpretation as a power/frequency distribution. If we wish to preserve this physical interpretation then it is clear that for non-stationary processes the "spectrum", however defined, must be allowed to become time-dependent. Consider, for example, a very simple non-stationary process which takes the form

$$X_t = \begin{cases} X_t^{(1)}, & t \le t_0, \\ X_t^{(2)}, & t > t_0, \end{cases}$$

where $\{X_t^{(1)}\}$, $\{X_t^{(2)}\}$, are each stationary processes but with different covariance structures. Assuming that the "change point" t_0 is known, the natural way to describe the power/frequency properties of $\{X_t\}$ is to introduce two spectra, one for the time interval $t \le t_0$, and the other for the interval $t > t_0$. For the more general model, $X_t = X_t^{(i)}$, $t_i \le t < t_{i+1}$, where the $\{X_t^{(i)}\}$ are different stationary processes, the properties of $\{X_t\}$ would be described by specifying a different form of spectrum for each of the intervals (t_i, t_{i+1}). Extending these ideas to the case of general types of non-stationary process, we are led to the notion of a *continuously changing spectrum*, i.e. a *time-dependent spectrum*. Clearly, in such cases we could never hope to estimate the spectrum at a particular instant of time, but if we assume that the spectrum is changing smoothly over time then, by using estimates which involve only local functions of the data, we may attempt to estimate some form of "average" spectrum of the process in the neighbourhood of any particular time instant. We therefore consider the class of non-stationary processes whose statistical characteristics are changing "smoothly" over time.

The notion of "frequency"

If we attempt to define a time-dependent spectrum which possesses a physical interpretation as a "local" power/frequency distribution, we must first decide what we mean by "frequency". This may appear to be a deceptively simple problem but its study is crucial. For, suppose we are given a real-valued continuous parameter non-stationary process, $X(t)$, and we have constructed some function, $p(\omega, t)$ say, such that, for each t,

$$E[X^2(t)] = \int_{-\infty}^{\infty} p(\omega, t)\, d\omega. \tag{6.2.1}$$

We certainly cannot conclude, on the basis of (6.2.1) alone, that $p(\omega, t)$ represents a decomposition of power over *frequency*, since in the above integral ω is merely a "dummy variable", and there is no reason why it should be related to the physical concept of "frequency". This point is highlighted by the following example. Let $R(s, t) = \text{cov}\{X(s), X(t)\}$ denote the covariance kernel of $\{X(t)\}$. Since $\{X(t)\}$ is a non-stationary process, $R(s, t)$ will no longer be a function of $|s-t|$ only, but $R(t-\tau/2, t+\tau/2)$ measures the covariance between values at time points separated by an interval τ and symmetrically placed about the time point t. If we try to interpret $R(t-\tau/2, t+\tau/2)$ as a form of "local autocovariance function", we may be tempted to define a time-dependent spectral density function by evaluating the Fourier transform of $R(t-\tau/2, t+\tau/2)$ with respect to

τ, keeping t fixed, leading to the function

$$\psi_t(\omega) = \frac{1}{2\pi} \int_{-\infty}^{\infty} R(t-\tau/2, t+\tau/2)\, e^{-i\omega\tau}\, d\tau. \tag{6.2.2}$$

This certainly produces a time-dependent function, and inverting (6.2.2) and setting $\tau = 0$ it is easily seen that

$$E[X^2(t)] = R(t, t) = \int_{-\infty}^{\infty} \psi_t(\omega)\, d\omega.$$

However, ψ_t certainly does not possess a physical interpretation as a power/frequency distribution since $\psi_t(\omega)$ may well take negative values. Despite the presence of the term $\{e^{i\omega t}\}$ in the integral in (6.2.2), the variable ω in $\psi_t(\omega)$ has no obvious physical interpretation as far as the properties of $\{X_t\}$ are concerned. (The function $\psi_t(\omega)$ is sometimes referred to as the *Wigner distribution*, and we will discuss it further in the following sections.)

Let us now return for a moment to the case of stationary processes. The reason why we can interpret the spectrum of a stationary process as a power/frequency distribution lies essentially in the fact that, when $X(t)$ is stationary, the process *itself* has a spectral representation of the form (cf. (1.3.12)),

$$X(t) = \int_{-\infty}^{\infty} e^{it\omega}\, dZ(\omega). \tag{6.2.3}$$

Heuristically, (6.2.3) tells us that any stationary process can be represented as the "sum" of sine and cosine waves with varying frequencies and (random) amplitudes and phases. We can now identify that *component* in $X(t)$ which has frequency ω, and meaningfully discuss the contribution of this *component* to the total power of the process. In the absence of such a representation we cannot immediately talk about "power distributions over frequency". But a non-stationary process cannot be represented as a "sum" of sine and cosine waves (with orthogonal coefficients)—instead, we have to represent it as a "sum" of other kinds of functions. Since, according to its conventional definition, the term "frequency" is associated specifically with sine and cosine functions, we cannot talk about the "frequency components" of a non-stationary process—unless we first define a more general concept of "frequency" which agrees with our physical understanding. The generalized notion of "frequency" on which the theory of *evolutionary spectra* is based may be seen from the following example. Suppose $X(t)$ is a deterministic function which has the form of a damped sine wave, say,

$$X(t) = A\, e^{-t^2/\alpha^2} \cos(\omega_0 t + \phi). \tag{6.2.4}$$

If we carry out a conventional Fourier analysis of $X(t)$ we see that it contains Fourier components at *all* frequencies—in fact, the Fourier transform of $X(t)$ consists of two Gaussian functions, one centred on ω_0 and the other on $-\omega_0$, the "width" of these functions being inversely proportional to the parameter α. In other words, if we represent $X(t)$ as a "sum" of sine and cosine functions with constant amplitudes, we need to include components at all frequencies. However, we can equally well describe $X(t)$ by saying that it consists of just two "frequency" components ($\omega = \omega_0$, $\omega = -\omega_0$), each component having a *time-varying amplitude*, $A\,e^{-t^2/\alpha^2}$. Indeed, if we were to examine the *local* behaviour of $X(t)$ in the neighbourhood of the time point t_0, this is precisely what we would observe, i.e. if the interval of observations was small compared with α, $X(t)$ would appear simply as a cosine function with frequency ω_0 and amplitude $A\,e^{-t_0^2/\alpha^2}$. However, it would not be physically meaningful to try to assign a "frequency" to a function $X(t)$ of arbitrary form; for example, it would make little sense to talk about the "frequency" of the function $\log \omega t$, and the parameter ω could certainly not be interpreted in this way. For the term "frequency" to be meaningful *the function $X(t)$ must possess what we can loosely describe as an "oscillatory form"*, and we can characterize this property by saying that the Fourier transform of such a function will be concentrated around a particular point ω_0 (or around $\pm\omega_0$ in the real case). Thus, if we have a non-periodic function $X(t)$ whose Fourier transform has an absolute maximum at the point ω_0, we may define ω_0 as *the frequency* of this function, the argument being that locally $X(t)$ behaves like a sine wave with (conventional) frequency ω_0, modulated by a "smoothly varying" amplitude function.

It is this type of reasoning which forms the basis of the "evolutionary spectra" approach, and, in fact, this approach rests on a spectral representation of non-stationary processes which is virtually a direct generalization of (6.2.4).

We may note, in passing, that the two different Fourier representations of the function (6.2.4) are equally valid, i.e. we may think of $X(t)$ as consisting either of an infinite number of frequency components with constant amplitudes, or as just two frequency components with time-varying amplitudes. These two representations correspond to the use of two different "families" of functions; in the former case the "family" consists of sines and cosines with constant amplitudes, and in the latter case it consists of sines and cosines with time-varying amplitude, $A\,e^{-t^2/\alpha^2}$. Unless we impose conditions on the form of the "family" of functions which we wish to use we have no way of expressing a preference for one or other of the two alternative representations.

A detailed account of the mathematical theory of evolutionary spectra is given in *Spec. Anal.*, Chapter 11, and it would not be appropriate to present a similar account in this chapter. Instead, we summarize in the next section the basic ideas underlying this theory, and in the following sections we describe some fields of application of evolutionary spectral analysis.

6.3 EVOLUTIONARY SPECTRA

Let $\{X(t), -\infty < t < \infty\}$ be a continuous parameter (complex valued) stochastic process, with zero mean and finite variance, i.e.

$$E[X(t)]=0, \qquad E[|X(t)|^2]<\infty, \quad \text{all } t.$$

The covariance kernel is then defined by

$$R(s, t) = E[X^*(s)X(t)]. \tag{6.3.1}$$

If $\{X(t)\}$ is stationary, i.e. if $R(s, t)$ is a function of $|s-t|$ only, then from Wiener-Khintchine theory (Section 1.3, equation (1.3.11)) $R(s, t)$ admits the representation

$$R(s, t) = \int_{-\infty}^{\infty} e^{i\omega(t-s)}\, dH(\omega), \tag{6.3.2}$$

where $H(\omega)$, the integrated spectrum, has the properties of a distribution function on the interval $(-\infty, \infty)$. Corresponding to the representation (6.3.2) for $R(s, t)$, $X(t)$ admits the spectral representation (cf. (1.3.12)),

$$X(t) = \int_{-\infty}^{\infty} e^{it\omega}\, dZ(\omega), \tag{6.3.3}$$

where $Z(\omega)$ is a process with orthogonal increments, and $E[|dZ(\omega)|^2] = dH(\omega)$. If we start from the representation (6.3.3) then (6.3.2) follows—so that $R(s, t)$ is a function of $|s-t|$ only and consequently $X(t)$ is stationary. It thus follows that if $X(t)$ is non-stationary it cannot be represented in the form (6.3.3) (with the $\{dZ(\omega)\}$ orthogonal), and similarly $R(s, t)$ cannot be represented in the form (6.3.2). However, we know from the theory of "general orthogonal expansions" (*Spec. Anal.*, p. 261) that for a fairly general class of stochastic processes $R(s, t)$ can be represented in a form similar to (6.3.2), provided we replace the functions $\{e^{i\omega t}\}$ by a more general "family" of functions, $\{\phi_t(\omega)\}$.

We now consider the class of processes for which *there exists a family $\mathcal{F}$ of functions $\{\phi_t(\omega)\}$ defined on the real line, and indexed by the suffix t, and a measure $\mu(\omega)$ on the real line*, such that for each s, t the covariance kernel $R(s, t)$ admits a representation of the form,

$$R(s, t) = \int_{-\infty}^{\infty} \phi_s^*(\omega)\phi_t(\omega)\, d\mu(\omega). \tag{6.3.4}$$

Note that although we have described $\mathscr{F}$ as a family of functions, each defined on the ω-axis and indexed by the parameter t, we may also think of $\mathscr{F}$ as a family of functions $\phi_\omega(t)$, say, each defined on the t-axis and indexed by the parameter ω. When we study the properties of various families, it is convenient to adopt the latter description.

In order for var$\{X(t)\}$ to be finite for each t, $\phi_t(\omega)$ must be quadratically integrable with respect to the measure μ, for each t. It then follows from the theory of general orthogonal expansions that whenever $R(s, t)$ has the representation (6.3.4), the process $\{X(t)\}$ admits a representation of the form,

$$X(t) = \int_{-\infty}^{\infty} \phi_t(\omega)\, dZ(\omega) \qquad (6.3.5)$$

where $Z(\omega)$ is an orthogonal process, with

$$E[|dZ(\omega)|^2] = d\mu(\omega).$$

The measure $\mu(\omega)$ here plays the same role as the integrated spectrum $H(\omega)$ in the case of stationary processes, so that the analogous situation to the case of an absolutely continuous spectrum is obtained by assuming that the measure $\mu(\omega)$ is absolutely continuous with respect to Lebesgue measure.

Parzen (1959) has pointed out that if there exists a representation of $\{X(t)\}$ of the form (6.3.5), then there is a multitude of different representations of the process, each representation based on a different family of functions. When the process is stationary, one valid choice of functions is the complex exponential family,

$$\phi_t(\omega) = e^{i\omega t}. \qquad (6.3.6)$$

This family provides the well-known spectral decomposition in terms of sine and cosine "waves", and forms the basis of the physical interpretation of spectral analysis as a "power distribution over frequency". If the process is non-stationary this choice of family of functions is no longer valid—as we would expect, since the sine and cosine waves are themselves "stationary" and it is natural that they should form the "basic elements" in building up models of stationary processes. If we wish to introduce the notion of frequency in the analysis of non-stationary processes, we must seek new "basic elements" which, although "non-stationary", have an *oscillatory form*, and in which *the notion of "frequency" is still dominant.* One class of basic elements (or more precisely, family of functions) which possess the required structure may be obtained as follows.

Suppose that, for each fixed ω, $\phi_t(\omega)$ (considered as a function of t) possesses a (generalized) Fourier transform whose modulus has an *absolute maximum* at frequency $\theta(\omega)$, say. Then we may regard $\phi_t(\omega)$ as an amplitude-modulated sine wave with frequency $\theta(\omega)$, and write $\phi_t(\omega)$ in the form

$$\phi_t(\omega) = A_t(\omega)\, e^{i\theta(\omega)t} \tag{6.3.7}$$

where the modulating function $A_t(\omega)$ is such that the modulus of its (generalized) Fourier transform has an absolute maximum at *the origin* (*i.e. zero frequency*).

The function of t, $\phi_t(\omega)$, will be called an *oscillatory function* if, for some $\theta(\omega)$ it may be written in the form (6.3.7) where $A_t(\omega)$ is of the form

$$A_t(\omega) = \int_{-\infty}^{\infty} e^{it\theta}\, dK_\omega(\theta) \tag{6.3.8}$$

with $|dK_\omega(\theta)|$ having an absolute maximum at $\theta = 0$. The function $A_t(\omega)$ may now be regarded as the "envelope" of $\phi_t(\omega)$. If, further, the family $\{\phi_t(\omega)\}$ is such that $\theta(\omega)$ is a single-valued function of ω (i.e. if no two distinct members of the family have Fourier transforms whose maxima occur at the same point), then we may transform the variable in the integral in (6.3.4) from ω to $\theta(\omega)$, and by suitably redefining $A_t(\omega)$ and the measure $\mu(\omega)$, write

$$R(s, t) = \int_{-\infty}^{\infty} A_s^*(\omega) A_t(\omega)\, e^{i\omega(t-s)}\, d\mu(\omega) \tag{6.3.9}$$

and correspondingly

$$X(t) = \int_{-\infty}^{\infty} A_t(\omega)\, e^{i\omega t}\, dZ(\omega) \tag{6.3.10}$$

where

$$E[|dZ(\omega)|^2] = d\mu(\omega).$$

If there exists a family of oscillatory functions $\{\phi_t(\omega)\}$ in terms of which the process $\{X(t)\}$ has a representation of the form (6.3.5), $\{X(t)\}$ will be called an *oscillatory process.*

It follows that any oscillatory process also has a representation of the form (6.3.10), where the family $A_t(\omega)$ satisfies the condition following (6.3.8) and that, w.l.o.g., we may write any family of oscillatory functions in the form

$$\phi_t(\omega) = A_t(\omega)\, e^{i\omega t}.$$

Since (6.3.6) is a particular case of (6.3.7) (with $A_t(\omega) \equiv 1$, all t, ω and $\theta(\omega) \equiv \omega$), the class of oscillatory processes certainly includes all second-order stationary processes.

Evolutionary spectra

Consider an oscillatory process of the form (6.3.10), with autocovariance kernel $R(s, t)$ of the form (6.3.9). For any particular process $\{X(t)\}$ there will, in general, be a large number of different families of oscillatory functions in terms of each of which $\{X(t)\}$ has a representation of the form (6.3.10), with each family inducing a different measure $\mu(\omega)$. For a particular family $\mathscr{F}$ of spectral functions $\{\phi_t(\omega)\}$ it is tempting to define the spectrum of $X(t)$ (with respect to $\mathscr{F}$) simply as the measure $\mu(\omega)$. However, such a definition would not have the interpretation of a "power distribution over frequency". For, from (6.3.9) we may write

$$\text{var}\{X(t)\} \equiv R(t, t) = \int_{-\infty}^{\infty} |A_t(\omega)|^2 \, d\mu(\omega). \tag{6.3.11}$$

Since var$\{X(t)\}$ may be interpreted as a measure of the "total power" of the process at time t, (6.3.11) gives a decomposition of total power in which the contribution from "frequency" ω is $\{|A_t(\omega)|^2 \, d\mu(\omega)\}$. This result is consistent with the interpretation of equation (6.3.10) as an expression for $X(t)$ as the limiting form of a "sum" of sine waves with different frequencies and time-varying random amplitudes $\{A_t(\omega) \, dZ(\omega)\}$. We are thus led to the following definition.

Let $\mathscr{F}$ denote a particular family of oscillatory functions, $\{\phi_t(\omega)\} \equiv \{A_t(\omega) \, e^{i\omega t}\}$, and let $\{X(t)\}$ be an oscillatory process having a representation of the form (6.3.10) in terms of the family $\mathscr{F}$. We define the *evolutionary power spectrum at time t with respect to the family* $\mathscr{F}$, $dH_t(\omega)$, by

$$dH_t^*(\omega) = |A_t(\omega)|^2 \, d\mu(\omega). \tag{6.3.12}$$

Note that when $\{X(t)\}$ is stationary, and $\mathscr{F}$ is chosen to be the family of complex exponentials, $dH_t(\omega)$ reduces to the standard definition of the (integrated) spectrum. The evolutionary spectrum has the same physical interpretation as the spectrum of a stationary process, namely that it describes a distribution of power over frequency, but whereas the latter is determined by the behaviour of the process over all time, the former represents specifically the spectral content of the process in the neighbourhood of the time instant t.

Although, according to the above definition the evolutionary spectrum, $dH_t(\omega)$, depends on the choice of family $\mathscr{F}$, it follows from (6.3.11) that

$$\text{var}\{X(t)\} = \int_{-\infty}^{\infty} dH_t(\omega) \tag{6.3.13}$$

so that the value of the integral of $dH_t(\omega)$ is independent of the particular family $\mathscr{F}$, and, for all families, represents the total power of the process at time t.

It is convenient to "standardize" the functions $A_t(\omega)$ so that, for all ω,

$$A_0(\omega) = 1,$$

i.e. we incorporate $|A_0(\omega)|$ in the measure $\mu(\omega)$. With this convention, $d\mu(\omega)$ represents the evolutionary spectrum at $t=0$, and $|A_t(\omega)|^2$ represents the change in the spectrum, relative to zero time. We now have, for each ω,

$$\int_{-\infty}^{\infty} dK_\omega(\theta) = 1,$$

so that the Fourier transforms of the $\{A_t(\omega)\}$ are normalized to have unit integrals.

When the measure $\mu(\omega)$ is absolutely continuous with respect to Lebesgue measure we may write for each t,

$$dH_t(\omega) = h_t(\omega)\, d\omega, \tag{6.3.14}$$

and $h_t(\omega)$ is then the *evolutionary spectral density function.*

Semi-stationary processes

As pointed out above, there usually exist a multitude of evolutionary spectral representations for a given oscillatory process, corresponding to all possible choices of the family $\mathscr{F}$. It is natural, therefore, to seek that family $\mathscr{F}^*$ for which the $\{A_t(\omega)\}$ are, in some sense, "most slowly varying". Since spectral estimation involves some form of averaging operation in the time domain, it is clear that the evolutionary spectra defined w.r. to $\mathscr{F}^*$ are those which we can most effectively estimate from data.

Suppose then that $\{X(t)\}$ is an oscillatory process whose non-stationary characteristics are changing "slowly" over time—in the sense that there exists a family $\mathscr{F}$ of oscillatory functions $\phi_t(\omega) = A_t(\omega)\, e^{i\omega t}$ in terms of which $\{X(t)\}$ has a representation of the form (6.3.10), and which are such that for each ω, $A_t(\omega)$ is, in some sense, a slowly varying function of t. In this context the most convenient characterization of a slowly varying function is obtained by specifying that its Fourier transform must be "highly concentrated" in the region of zero frequency.

For each family $\mathscr{F}$, we define the function $B_{\mathscr{F}}(\omega)$ by

$$B_{\mathscr{F}}(\omega) = \int_{-\infty}^{\infty} |\theta|\,|dK_\omega(\theta)|, \tag{6.3.15}$$

which is a measure of the "width" of $|dK_\omega(\theta)|$.

A family $\mathscr{F}$ of oscillatory functions is called *semi-stationary* if the function $B_{\mathscr{F}}(\omega)$ is bounded for all ω, and the constant B defined by

$$B_{\mathscr{F}} = \sup_{\omega}\{B_{\mathscr{F}}(\omega)\}^{-1} \tag{6.3.16}$$

is called the *characteristic width* of the family $\mathscr{F}$.

A *semi-stationary process* $\{X(t)\}$ is now defined as one for which there exists a semi-stationary family $\mathcal{F}$ in terms of which $X(t)$ has a representation of the form (6.3.10).

For example, the *uniformly modulated process*, $X(t)=C(t)X_t^{(0)}$, where $X_t^{(0)}$ is stationary and $C(t)$ is a deterministic function, is a semi-stationary process, since the family $\mathcal{F}_0 \equiv \{C(t)\, e^{i\omega t}\}$ is semi-stationary. (Note that, since $C(t)$ is independent of ω, $B_{\mathcal{F}_0}(\omega)$ is independent of ω.)

For a particular semi-stationary process $\{X(t)\}$ consider the class $\mathscr{C}$ of semi-stationary families $\mathcal{F}$, in terms of each of which $\{X(t)\}$ admits a spectral representation. We define the *characteristic width of the process* $\{X(t)\}$, B_X, by

$$B_X = \sup_{\mathcal{F} \in \mathscr{C}} \{B_{\mathcal{F}}\}. \tag{6.3.17}$$

Roughly speaking, $2\pi B_X$ may be interpreted as the maximum interval over which the process may be treated as "approximately stationary". For stationary processes the class $\mathscr{C}$ contains the family of complex exponentials, which has infinite characteristic width. Consequently, all stationary processes have infinite characteristic width.

Now let $\mathscr{C}^* \subset \mathscr{C}$ denote the sub-class of families whose characteristic widths are each equal to B_X, and let $\mathcal{F}^*$ denote any family $\in \mathscr{C}^*$. For example, if $\{X(t)\}$ is stationary, $\mathscr{C}^*$ contains only one family, namely the complex exponentials, so that $\mathcal{F}^*$ is uniquely determined as this family. (However, as far as the theory of evolutionary spectra is concerned, the uniqueness of $\mathcal{F}^*$ is not required.) If $\mathscr{C}^*$ is empty, let $\mathcal{F}^*$ denote any family whose characteristic width is arbitrarily close to B_X.

We now consider the spectral representation of $\{X(t)\}$ in terms of the family $\mathcal{F}^*$. Thus we write

$$X(t) = \int_{-\infty}^{\infty} A_t^*(\omega)\, e^{i\omega t}\, dZ^*(\omega) \tag{6.3.18}$$

where $E[|dZ^*(\omega)|^2] = d\mu^*(\omega)$, say, and the functions $\phi_t(\omega) = \{A_t^*(\omega)\, e^{i\omega t}\} \in \mathcal{F}^*$. It may then be shown (*Spec. Anal.*, Chapter 11) that the evolutionary spectra defined w.r. to $\mathcal{F}^*$ can be efficiently estimated by linear filtering techniques provided the "width" of the filter impulse response function, $\{g(u)\}$, defined by,

$$B_g = \int_{-\infty}^{\infty} |u||g(u)|\, du \tag{6.3.19}$$

is much smaller than B_X. However, the width of the impulse response function affects also the degree of resolvability which we can attain in the frequency domain: for high frequency domain resolution we require $\{g(u)\}$

to decay "slowly", i.e. B_g should be large, but for high time domain resolution we require $\{g(u)\}$ to decay "quickly", i.e. B_g should be small.

In other words, the more accurately we try to determine $dH_t(\omega)$ as a function of time, the less accurately we determine it as a function of frequency, and vice versa. This feature leads to a form of UNCERTAINTY PRINCIPLE, namely *in determining evolutionary spectra, one cannot obtain simultaneously a high degree of resolution in both the time domain and frequency domain* (Priestley, 1965). Daniels (1965) has pointed out that this uncertainty principle is completely analogous to Heisenberg's uncertainty principle in quantum mechanics. (The analogy with quantum mechanical concepts is discussed further by Tjøstheim (1976).)

Suppose now that we fix the degree of resolution in the frequency domain, i.e. we set a lower bound to B_g. For a particular family $\mathscr{F}$, the resolution in the time domain will be determined by the value of $B_g/B_{\mathscr{F}}$. Clearly then, we obtain the maximum possible resolution in time by working in terms of the family with the maximum characteristic width. Thus, if $\mathscr{C}^*$ contains only one member, i.e. if $\mathscr{F}^*$ is uniquely determined, then $\mathscr{F}^*$ provides the *natural representation* for $\{X(t)\}$, and is the family in terms of which we can most precisely express the time-varying spectral pattern of the process. In particular, we now see why the natural representation of stationary processes is given in terms of the complex exponential family—the reason being simply that in this case $\mathscr{F}^*$ is unique and is just this family.

Discrete parameter processes

The above theory may be extended in a fairly obvious way to the case of discrete parameter non-stationary processes. Thus, an oscillatory discrete parameter process will have a representation of the form

$$X_t = \int_{-\pi}^{\pi} e^{it\omega} A_t(\omega)\, dZ(\omega), \qquad t = 0, \pm 1, \pm 2, \ldots, \tag{6.3.20}$$

where, for each ω, the sequence $\{A_t(\omega)\}$ has a generalized Fourier transform whose modulus has an absolute maximum at origin, and $Z(\omega)$ is an orthogonal process on the interval $(-\pi, \pi)$, with $E[|dZ(\omega)|^2] = d\mu(\omega)$.

The evolutionary spectrum at time t with respect to the family of sequences $\mathscr{F} \equiv \{e^{it\omega} A_t(\omega)\}$ is then defined as,

$$dH_t(\omega) = |A_t(\omega)|^2\, d\mu(\omega), \qquad -\pi \le \omega \le \pi. \tag{6.3.21}$$

When the measure $\mu(\omega)$ is absolutely continuous we may again write

$$dH_t(\omega) = h_t(\omega)\, d\omega,$$

where $h_t(\omega)$, the evolutionary spectral density function, exists for all ω in $(-\pi, \pi)$.

Estimation of evolutionary spectra

Suppose we are given observations over the interval $(0, T)$ from a continuous parameter semi-stationary process whose evolutionary spectra (w.r. to $\mathscr{F}^*$) are absolutely continuous with evolutionary spectral density functions $\{h_t(\omega)\}$. The estimation of $h_t(\omega)$ is performed in two stages: (i) we first pass the data through a linear filter centred on frequency ω_0, say, yielding an output $U(t)$, say; (ii) we then compute a weighted average of $|U(t)|^2$ in the neighbourhood of the time point t to provide an estimate of the local power density at frequency ω_0. Thus, given observations $\{X(t)\}$, $0 \le t \le T$, we set

$$U(t) = \int_{t-T}^{t} g(u) X(t-u)\, e^{-i\omega_0(t-u)}\, du, \tag{6.3.22}$$

$$\hat{h}_t(\omega_0) = \int_{t-T}^{t} w(v) |U(t-v)|^2\, dv. \tag{6.3.22a}$$

Here, $\{g(u)\}$ is a filter whose transfer function

$$\Gamma(\omega) = \int_{-\infty}^{\infty} g(u)\, e^{-i\omega u}\, du$$

is peaked in the neighbourhood of $\omega = 0$, is normalized so that $\int_{-\infty}^{\infty} |\Gamma(\omega)|^2\, d\omega = 1$, and the filter "width" B_g is much smaller than B_X. The function $w(v)$ is chosen so that its "width" is much larger than B_g, and is normalized so that $\int_{-\infty}^{\infty} w(v)\, dv = 1$. We assume that both $g(u)$ and $w(v)$ decay to zero sufficiently fast so that the limits in the integrals in (6.3.22), (6.3.22a) may be replaced effectively by $(-\infty, \infty)$. It may then be shown (Priestley, 1966) that

$$E[\hat{h}_t(\omega_0)] \sim \int_{-\infty}^{\infty} \bar{h}_t(\omega + \omega_0) |\Gamma(\omega)|^2\, d\omega \tag{6.3.23}$$

where

$$\bar{h}_t(\omega + \omega_0) = \int_{-\infty}^{\infty} w(v) h_{t-v}(\omega + \omega_0)\, dv.$$

Making the usual assumption that $h_t(\omega)$ is "flat" over the bandwidth of $|\Gamma(\omega)|^2$, we may write

$$E[\hat{h}_t(\omega_0)] \sim \bar{h}_t(\omega_0). \tag{6.3.23a}$$

Thus, $\hat{h}_t(\omega_0)$ is an approximately unbiased estimate of the weighted average of $h_t(\omega_0)$ in the neighbourhood of the time point t. It may be shown further that

$$\text{var}[\hat{h}_t(\omega_0)] \sim \tilde{h}_t^2(\omega_0)\left\{\int_{-\infty}^{\infty} |W(\omega)|^2\, d\omega\right\}\left\{\int_{-\infty}^{\infty} |\Gamma(\omega)|^4\, d\omega\right\}, \qquad \omega_0 \neq 0 \tag{6.3.23b}$$

where

$$\tilde{h}_t^2(\omega_0) = \frac{\int_{-\infty}^{\infty} h_{t-v}^2(\omega)\{w(v)\}^2\, dv}{\int_{-\infty}^{\infty} \{w(v)\}^2\, dv}$$

and

$$W(\omega) = \int_{-\infty}^{\infty} w(v)\, e^{-i\omega v}\, dv.$$

For a discrete parameter process, $\{X_t\}$, the expressions corresponding to (6.3.22), (6.3.22a), become

$$U_t = \sum_u g_u X_{t-u}\, e^{-i\omega_0(t-u)},$$

$$\hat{h}_t(\omega_0) = \sum_v w_v |U_{t-v}|^2,$$

where now $\{g_u\}$, $\{w_v\}$ are suitably chosen sequences. The mean and variance of $\hat{h}_t(\omega_0)$ are given by making the usual modifications to (6.3.23), (6.3.23a), i.e. the limits of the integrals in (6.3.23), (6.3.23a) become $(-\pi, \pi)$, and the integrals defining $\Gamma(\omega)$, $W(\omega)$, $\bar{h}_t(\omega_0)$, and $\tilde{h}_t^2(\omega)$, are replaced by sums.

A detailed derivation of the above results is given in Priestley (1966). To implement this estimation procedure we need to specify the "windows" $\{g(u)\}$, $\{w(v)\}$. The choice of these windows is discussed in Priestley (1966), and some particular examples are given in Priestley (1965). Some numerical examples of evolutionary spectral analysis are presented in *Spec. Anal.*, Ch. 11.

An extension of the theory of evolutionary spectra to the case of processes with discrete spectra is discussed by Martin (1981).

6.3.1 Evolutionary Cross-spectra

An extension of the theory of evolutionary spectra to the case of bivariate non-stationary processes was developed by Priestley and Tong (1973). The basic ideas may be described as follows. Consider, for example, a continuous parameter bivariate process. $\{X(t), Y(t)\}$ in which each component is an oscillatory process. Then with an obvious notation, we can write

$$X(t) = \int_{-\infty}^{\infty} A_{t,X}(\omega)\, e^{i\omega t}\, dZ_X(\omega), \tag{6.3.24}$$

and

$$Y(t) = \int_{-\infty}^{\infty} A_{t,Y}(\omega)\, e^{i\omega t}\, dZ_Y(\omega), \tag{6.3.25}$$

where

$$\left.\begin{aligned} E[dZ_X(\omega)\, dZ_X^*(\omega')] &= E[dZ_Y(\omega)\, dZ_Y^*(\omega')] \\ &= E[dZ_X(\omega)\, dZ_Y^*(\omega')] = 0 \qquad \text{for } \omega \neq \omega' \\ E[|dZ_X(\omega)|^2] &= d\mu_{XX}(\omega),\; E[|dZ_Y(\omega)|^2] = d\mu_{YY}(\omega), \text{ and} \\ E[dZ_X(\omega)\, dZ_Y^*(\omega)] &= d\mu_{XY}(\omega). \end{aligned}\right\}$$

Let $\mathscr{F}_X$, $\mathscr{F}_Y$ denote respectively the families of oscillatory functions $\{\phi_{t,X}(\omega) \equiv A_{t,X}(\omega)\, e^{i\omega t}\}$, $\{\phi_{t,Y}(\omega) \equiv A_{t,Y}(\omega)\, e^{i\omega t}\}$. We define the *evolutionary power cross-spectrum at time t with respect to the families $\mathscr{F}_X$ and $\mathscr{F}_Y$*, $dH_{t,XY}(\omega)$, by

$$dH_{t,XY}(\omega) = A_{t,X}(\omega) A_{t,Y}^*(\omega)\, d\mu_{XY}(\omega). \tag{6.3.26}$$

Note that when we may choose $\mathscr{F}_X \equiv \mathscr{F}_Y$, equation (6.3.26) takes on the special form

$$dH_{t,XY}(\omega) = |A_t(\omega)|^2\, d\mu_{XY}(\omega). \tag{6.3.27}$$

where $A_t(\omega) \equiv A_{t,X}(\omega) \equiv A_{t,Y}(\omega)$. (Abdrabbo and Priestley (1969) used (6.3.27) as their definition of evolutionary cross-spectrum for this special case.) Further, if $\{X(t), Y(t)\}$ is a bivariate stationary process, so that $\mathscr{F}_X$ and $\mathscr{F}_Y$ may be chosen to be the family of complex exponentials, namely $\mathscr{F}_X \equiv \mathscr{F}_Y \equiv \{e^{i\omega t}\}$, $dH_{t,XY}(\omega)$ reduces to the classical definition of the cross-spectrum. Since, for each t, we may write

$$dH_{t,XY}(\omega) = E[A_{t,X}(\omega)\, dZ_X(\omega) A_{t,Y}^*(\omega)\, dZ_Y^*(\omega)]$$

it follows that $dH_{t,XY}(\omega)$ possesses a physical interpretation similar to that of the cross-spectrum of a bivariate stationary process, namely that it represents the average value of the product of the "amplitudes" of the corresponding frequency components in the two processes. Note, however, that in the non-stationary case these "amplitudes" are time-dependent so that, correspondingly, the cross-spectrum is also time-dependent.

Clearly, $dH_{t,XY}(\omega)$ is complex-valued, and in virtue of the Cauchy-Schwarz inequality we have immediately that,

$$|dH_{t,XY}(\omega)|^2 \leq dH_{t,XX}(\omega)\, dH_{t,YY}(\omega), \quad \text{all } t \text{ and } \omega. \tag{6.3.28}$$

If the measure $\mu_{XY}(\omega)$ is absolutely continuous with respect to Lebesgue measure, we may write, for each t,

$$dH_{t,XY}(\omega) = h_{t,XY}(\omega)\, d\omega, \tag{6.3.29}$$

and $h_{t,XY}(\omega)$ may then be termed the *evolutionary cross-spectral density function.* We may now write

$$h_{t,XY}(\omega) = c_{t,XY}(\omega) - iq_{t,XY}(\omega), \tag{6.3.30}$$

and term the real-valued functions $c_{t,XY}(\omega)$ and $q_{t,XY}(\omega)$ the *evolutionary co-spectrum* and the *evolutionary quadrature spectrum*, respectively. If the measures $\mu_{XX}(\omega)$ and $\mu_{YY}(\omega)$ are absolutely continuous, we may similarly define the *evolutionary auto-spectral density functions*, $h_{t,XX}(\omega)$ and $h_{t,YY}(\omega)$ (cf (6.3.14)). The function, $W_{YX}(\omega)$, defined by

$$W_{YX}(\omega) = |h_{t,YX}(\omega)| / \{h_{t,XX}(\omega) h_{t,YY}(\omega)\}^{1/2} \tag{6.3.31}$$

is termed the *coherency (spectrum)* between $\{X(t)\}$ and $\{Y(t)\}$. Equivalently, (6.3.31) may be written,

$$W_{YX}(\omega) = |E[dZ_Y(\omega)\, dZ_X^*(\omega)]| / \{E|dZ_X(\omega)|^2 E|\, dZ_Y(\omega)|^2\}^{1/2}. \tag{6.3.32}$$

We may thus interpret $W_{XY}(\omega)$ as the modulus of the correlation coefficient between $dZ_X(\omega)$, $dZ_Y(\omega)$, or more generally, as a measure of the *linear* relationship between the corresponding components at frequency ω in the processes $\{Y(t)\}$ and $\{X(t)\}$. As we can see from (6.3.32) this measure is *independent* of the time parameter t. On specializing to stationary processes, the above definition of coherency coincides with the standard definition of coherency. We note that, in virtue of (6.3.28) $0 \le W_{YX}(\omega) \le 1$ and $W_{XY}(\omega) = W_{YX}(\omega)$, all ω.

Suppose we are given a sample record of a continuous parameter process, $\{X(t), Y(t)\}$, say for $t \in [0, T_0]$. One method of estimating the evolutionary cross-spectral density function, $h_{t,XY}(\omega)$, is based on an extension of the method previously described for dealing with the univariate case. We may briefly describe the method as follows.

Let $\{g(u)\}$ be a filter satisfying the usual conditions with transfer function

$$\Gamma(\omega) = \int_{-\infty}^{\infty} g(u)\, e^{-iu\omega}\, du$$

and width

$$B_g = \int_{-\infty}^{\infty} |u||g(u)|\, du.$$

We choose $\{g(u)\}$ so that, $B_g \ll \min(B_X, B_Y) \ll T_0$, and write, for any frequency ω_0,

$$U_X(t, \omega_0) = \int_{t-T_0}^{t} g(u)X(t-u)\, e^{-i\omega_0(t-u)}\, du$$

$$\doteqdot \int_{-\infty}^{\infty} g(u)X(t-u)\, e^{-i\omega_0(t-u)}\, du,$$

$$U_Y(t, \omega_0) = \int_{t-T_0}^{t} g(u)Y(t-u)\, e^{-i\omega_0(t-u)}\, du$$

$$\doteqdot \int_{-\infty}^{\infty} g(u)Y(t-u)\, e^{-i\omega_0(t-u)}\, du.$$

Now let $w(t)$ be a weight-function satisfying the usual conditions (cf. Priestley, 1965). Then we may estimate $h_{t,XY}(\omega_0)$ by

$$\hat{h}_{t,XY}(\omega_0) = \int_{t-T_0}^{t} w(u)[U_X(t-u, \omega_0)U_Y^*(t-u, \omega_0)]\, du. \qquad (6.3.33)$$

The weight function $w(u)$ is usually chosen so that its effective "width" is *much larger than* B_g, the width of $\{g(u)\}$, and so that it decays to zero sufficiently fast for the limits of the integral in (6.3.33) to be replaced by $(-\infty, \infty)$. We then obtain,

$$E[\hat{h}_{t,XY}(\omega_0)] \sim \bar{h}_{t,XY}(\omega_0), \qquad (6.3.34)$$

where

$$\bar{h}_{t,XY}(\omega) = \int_{-\infty}^{\infty} w(u)h_{t-u,XY}(\omega)\, du. \qquad (6.3.35)$$

Thus, $\hat{h}_{t,XY}(\omega_0)$ is an approximately unbiased estimate of the weighted average value of $h_{t,XY}(\omega_0)$ in the neighbourhood of the time point t.

An investigation of the sampling properties of $\hat{h}_{t,XY}(\omega)$ is given in Priestley and Tong (1973). The method of investigation is similar to that of Priestley (1966), but the calculations and results are much more lengthy. We summarize here only the main results. Making the usual assumptions concerning the "width" of the weight function $w(u)$ (see, for example, Priestley, 1965, pp. 219–20), we may show that the covariance between $\hat{h}_{t,XY}(\omega)$ and $\hat{h}_{s,XY}(\omega')$ is effectively zero if either

(i) $|\omega \pm \omega'|$ is sufficiently large such that $|\omega \pm \omega'| \gg$ bandwidth of $|\Gamma(\omega)|^2$, or
(ii) $|s-t|$ is sufficiently large, i.e. $|s-t| \gg$ "width" of the weight function $\{w(u)\}$.

We may further show that the variance–covariance matrix of $\{\hat{h}_{t,XX}(\omega_0), \hat{h}_{t,YY}(\omega_0), \hat{c}_{t,XY}(\omega_0), \hat{q}_{t,XY}(\omega_0)\}$ has a form similar to that arising in the case of bivariate stationary process (see, for example, *Spec. Anal.*, p. 702). Its detailed form is given in Subba Rao and Tong (1972).

Estimation of the transfer function of a time-dependent open loop system

In many practical applications the processes $\{X(t)\}$ and $\{Y(t)\}$ may be thought of as the "input" and "output" of an "open loop system", with the associated "block diagram" given in Fig. 6.1 (see, for example, *Spec. Anal.*, p. 786). Typically, this situation would be described by a linear time-invariant model of the form,

$$Y(t) = \int_{-\infty}^{\infty} d(u)X(t-u)\,du + e(t), \tag{6.3.36}$$

where $\{X(t)\}$, $\{Y(t)\}$, and $\{e(t)\}$ are stationary processes. However, a more general description is given by a time-dependent model of the form

$$Y(t) = \int_{-\infty}^{\infty} d_t(u)X(t-u)\,du + e(t), \tag{6.3.37}$$

where now $\{X(t)\}$, $\{Y(t)\}$, and $e(t)$ are semi-stationary processes and $\{d_t(u)\}$ describes a "slowly changing" filter. Suppose that we have available operating records of $\{X(t)\}$ and $\{Y(t)\}$ over the interval $0 \le t \le T$, and wish to estimate for $0 \le t \le T$ the time dependent transfer function,

$$D_t(\omega) = \int_{-\infty}^{\infty} d_t(u)\,e^{iu\omega}\,du. \tag{6.3.38}$$

Generalizing the well-known result for the time-invariant case (*Spec. Anal.*, p. 786), a natural estimate of $D_t(\omega)$ is given by

$$\hat{D}_t(\omega) = \hat{h}_{t,YX}(\omega)/\hat{h}_{t,XX}(\omega), \qquad \text{provided } \hat{h}_{t,XX}(\omega) \ne 0, \tag{6.3.39}$$

where $\hat{h}_{t,YX}$, $\hat{h}_{t,XX}$ denote the estimated cross- and auto-evolutionary spectral

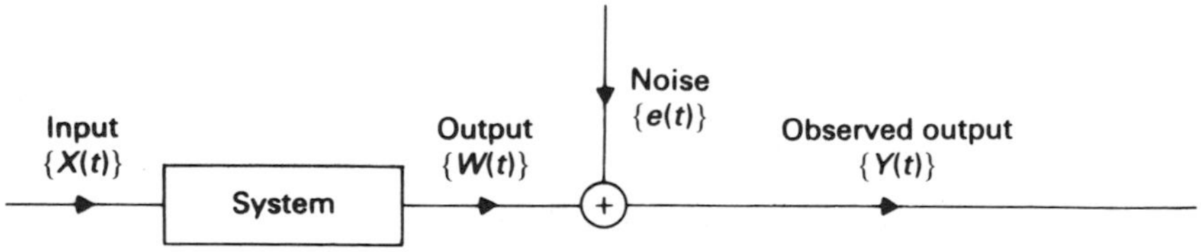

Fig. 6.1. Block diagram showing the processes $\{X(t)\}$ and $\{Y(t)\}$.

density functions computed from the given sample records. If we write

$$D_t(\omega) = G_t(\omega)\, e^{-i\phi_t(\omega)}, \tag{6.3.40}$$

where $G_t(\omega)$ and $\phi_t(\omega)$ may be termed respectively the *evolutionary gain-spectrum* and the *evolutionary phase-spectrum*, we may estimate $G_t(\omega)$ and $\phi_t(\omega)$ respectively by

$$\hat{G}_t(\omega) = \{\hat{c}^2_{t,YX}(\omega) + \hat{q}^2_{t,YX}(\omega)\}^{1/2} / \hat{h}_{t,XX}(\omega), \tag{6.3.41}$$

and

$$\hat{\phi}_t(\omega) = \tan^{-1}\{\hat{q}_{t,YX}(\omega) / \hat{c}_{t,YX}(\omega)\}. \tag{6.3.42}$$

The sampling properties of $\hat{G}_t(\omega)$, $\hat{\phi}_t(\omega)$ are discussed in Priestley and Tong (1973), who also give some numerical examples of the methods described in this section.

Tests for the time-dependence of linear open-loop systems are given in Subba Rao and Tong (1972, 1973).

6.4 OTHER DEFINITIONS OF TIME-DEPENDENT "SPECTRA"

In addition to the theory of evolutionary spectra, various other definitions of time-dependent "spectra" for non-stationary processes have been proposed. These may be summarized as follows.

1. *Page's instantaneous power spectra*

The first attempt to define time-dependent spectra is due to Page (1952), who, in a pioneering paper, introduced the term *instantaneous power spectra.* Given a continuous parameter process $\{X(t)\}$, Page first introduces the quantity

$$g^*_T(\omega) = \left| \int_0^T X(t)\, e^{-i\omega t}\, dt \right|^2, \tag{6.4.1}$$

($g^*_T(\omega)$ is, in fact, proportional to the periodogram of the observations on the interval $(0, T)$) and defines the "spectrum" $f^*(\omega)$ of the process as

$$f^*(\omega) = \lim_{T\to\infty} E\{g^*_T(\omega)\}. \tag{6.4.2}$$

Page then defines the "instantaneous power spectrum", $\rho_t(\omega)$, by writing

$$E\{g^*_T(\omega)\} = \int_0^T \rho_t(\omega)\, dt,$$

so that

$$\rho_t(\omega) = \frac{d}{dt} E\{g^*_t(\omega)\}, \tag{6.4.3}$$

and

$$f^*(\omega) = \int_0^\infty \rho_t(\omega)\, dt.$$

The instantaneous power spectrum $\{\rho_t(\omega)\, dt\}$ represents, roughly speaking, the difference between the power distributions of the process over the interval $(0, t)$ and over the interval $(0, t+dt)$. That is, if we take one record of $X(t)$ over the interval $(0, t)$ and another record over the interval $(0, t+dt)$, pass each through an ideal wave-analyser, and measure the difference between these two power spectra, then the expected value of this difference at frequency ω corresponds to $\{\rho_t(\omega)\, dt\}$. However, this is, in general, quite different from the evolutionary spectrum which represents the power distribution of the process *within* the interval $(t, t+dt)$, i.e. the evolutionary spectrum corresponds roughly to what we would measure if we took just that portion of $X(t)$ lying between t and $t+dt$ and passed it through a wave-analyser. (In the stationary case both definitions reduce to the classical definition of the spectrum.) From the point of view of physical interpretation, the evolutionary spectrum would seem to be the more relevant quantity.

2. *The Wigner spectrum*

Let $X(t)$ be a continuous parameter non stationary process, with covariance kernel $R(s, t)$. The Wigner spectrum, $\psi_t(\omega)$, is defined as the Fourier transform of $R(t-\tau/2, t+\tau/2)$, regarded as a function of τ with t fixed. Thus, $\psi_t(\omega)$ is given by (cf. (6.2.2)),

$$\psi_t(\omega) = \frac{1}{2\pi} \int_{-\infty}^{\infty} R(t-\tau/2, t+\tau/2)\, e^{-i\omega\tau}\, d\tau. \tag{6.4.4}$$

The function $\psi_t(\omega)$ clearly reduces to the classical definition of the spectrum when the process is stationary, and its mathematical form bears a superficial resemblance to the classical definition. However, as previously noted, $\psi_t(\omega)$ lacks completely a meaningful physical interpretation. This definition of a time-dependent spectrum was discussed by Mark (1970), who pointed out that, for example, $\psi_t(\omega)$ may take negative values for certain processes. Mark therefore introduced a second type of "spectrum", called the *physical spectrum*, $S(\omega, t, W)$, defined as follows. Let $W(t)$ be a suitable real-valued function so that $W(0)>0$, $W(t)$ is "small" outside the neighbourhood to $t=0$, and $\int_{-\infty}^{\infty} W^2(t)\, dt = 1$. Then $S(\omega, t, W)$ is defined by

$$S(\omega, t, W) = E\left[\left|\int_{-\infty}^{\infty} W(t-u)X(u)\, e^{-i\omega u}\, du\right|^2\right]. \tag{6.4.5}$$

Unfortunately, $S(\omega, t, W)$ is of little use as a *definition* of a time-dependent spectrum since it involves an arbitrary function $W(t)$. Thus, each possible choice of $W(t)$ leads to a different expression for $S(\omega, t, W)$, and no particular $S(\omega, t, W)$ is characteristic purely of the process $X(t)$. However, suppose we consider an "idealized" form of $S(\omega, t, W)$ in which the function $W(t)$ becomes in the limit a δ-function. In this case $S(\omega, t, W)$ would describe the behaviour of $X(t)$ at a single time point and would not be "contaminated" by the behaviour of $X(t)$ at other time points. But it is impossible to achieve this type of "idealized" spectrum by considering a limiting form of $S(\omega, t, W)$ since, as $W(t) \to \delta(t)$, the transfer function of the filter corresponding to $W(t)$ becomes uniform, and $S(\omega, t, W) \to E[X^2(t)]$ and so has lost all dependence on the frequency variable ω. On the other hand, if we try to design a filter which has perfect frequency selectivity, then its transfer function must have zero bandwidth and consequently its impulse response function will never die out. This, in turn, means that the corresponding expression for $S(\omega, t, W)$ will now depend on the behaviour of $X(t)$ over all time, and will not, therefore, describe the spectral properties of the process in the neighbourhood of the specific time point t. In other words, $S(\omega, t, W)$ will have lost all form of time dependence. This feature is, of course, a consequence of the basic *uncertainty principle* which tells us that we cannot obtain simultaneously an arbitrarily high degree of resolution in both the time and frequency domains.

Nevertheless, if $W(t)$ is not chosen to correspond to either of the two extreme forms, an expression of the form (6.3.5) (modified so that the expectation is replaced by a "time domain" average) can provide a very useful *estimate* of the "idealized" spectrum. In fact, if we identify the function $W(t)$ with the function $g(u)$ in (6.3.22), then we see at once that $S(\omega, t, W)$ *is simply the expression* $E[|U(t)|^2]$ *and by* (6.3.23) *this is just a "smoothed" version of the evolutionary spectral density function,* $h_t(\omega)$. Moreover, if we replace the expectation in (6.4.5) by a time domain average, then $S(\omega, t, W)$ becomes identical with the evolutionary spectral estimate as given by (6.3.22a).

3. *Tjøstheim's approach*

Tjøstheim (1976) proposed a definition of a time-dependent spectrum based on the representation of a process in terms of its "innovations". Given a discrete parameter non-stationary process $\{X_t\}$, Cramer (1961a) has shown that if $\{X_t\}$ is "purely non-deterministic" (cf. *Spec. Anal.*, p. 755) it possesses a one-sided linear representation of the form

$$X_t = \sum_{u=0}^{\infty} a_t(u)\varepsilon_{t-u}, \tag{6.4.6}$$

where $\{\varepsilon_t\}$ is a stationary white noise process, with variance σ_ε^2, say. (Note that since X_t is non-stationary with coefficients, $a_t(u)$, in (6.4.6) will be time-dependent.) Tjøstheim then defines the "spectrum at time t" by

$$p_t(\omega) = \frac{\sigma_\varepsilon^2}{2\pi} \left| \sum_{u=0}^{\infty} a_t(u)\, e^{-i\omega u} \right|^2. \tag{6.4.7}$$

Comparing (6.4.6) with the discrete parameter form of equation (11.2.15) of *Spec. Anal.* (p. 826), we see that $p_t(\omega)$ *is, in fact, exactly the same as the evolutionary spectrum of* $\{X_t\}$ *defined with respect to the family* $\{A_t^{(1)}(\omega) = \sum_{u=0}^{\infty} a_t(u)\, e^{-i\omega u}\}$ (and with $d\mu(\omega) \propto d\omega$). Thus, Tjøstheim's definition is a special case of the evolutionary spectrum definition, the emphasis being placed on the family of functions generated by the coefficients in (6.4.6). This approach leads to a uniquely defined, time-dependent spectrum, and is certainly attractive from a theoretical point of view. However, the major drawback with this definition lies in the fact that unless the family $\{A_t^{(1)}(\omega)\}$ is *oscillatory* the variable ω in (6.4.7) will not possess a physical interpretation in terms of *frequency*, and consequently $p_t(\omega)$ will not represent a "power/frequency" distribution. Also, the estimation procedure described in Section 6.3 leads inevitably to the evolutionary spectrum defined with respect to the family $\mathscr{F}^*$ of maximum characteristic width. It is difficult to see, therefore, how one could estimate $p_t(\omega)$ empirically from quadratic functions of the observations—unless, of course, one had prior knowledge of the form of the $\{a_t(u)\}$ sequence.

4. Melard's approach

In an interesting and extensive series of papers, Melard and his co-workers have made important contributions to the study of spectral analysis of non-stationary processes—see, for example, Melard (1975, 1978, 1985a, 1985b), Kiehm and Melard (1981), Melard and Wybouw (1984), and De Schutter-Herteleer and Melard (1986). The basic idea underlying Melard's approach is to use the time-dependent linear representation of a process in terms of its innovations, i.e. the representation given by (6.4.6). He then uses the coefficients in this representation to construct the time-dependent "spectrum" given by (6.4.7)—as in Tjøstheim's approach. (This definition was proposed independently by Melard (1975) and Tjøstheim (1976). Melard describes these time-dependent functions as "*evolutive spectra*".) Melard's approach thus shares the same features as Tjøstheim's, namely that it will not necessarily possess a physical interpretation as a local power/frequency distribution but may provide a useful mathematical characterization of the probabilistic properties of the process.

The loss of the physical interpretation is due to the fact that the family $A_t^{(1)}(\omega)=\sum_{u=0}^{\infty} a_t(u)\, e^{-i\omega u}$ (the $\{a_t(u)\}$ being the coefficients in linear representation (6.4.6)) is not necessarily oscillatory, i.e. does not necessarily satisfy the condition stated below equation (6.3.8). However, Melard points out that the relaxation of this condition simplifies considerably the mathematical properties of the resulting time-dependent spectrum. Indeed, the problem of finding spectral representations in terms of families of oscillatory functions, and, more especially, the selection of the family $\mathscr{F}^*$ of maximum characteristic width, can involve complicated mathematical analyses (except, of course, for the stationary case where $\mathscr{F}^*$ is easily seen to be the family of complex exponentials). Melard (1985b) is one of the few papers which studies these problems, and although he considers only fairly simple non-stationary structures (such as processes which take non-zero values only at two points), his analysis is a significant contribution to the mathematical development of evolutionary spectral analysis.

The paper by De Schutter-Herteleer and Melard (1986) develops an extension of Melard's "evolutive spectrum" to the case of bivariate non-stationary processes. The basic ideas are similar to those described in Section 6.3.1 (with the families $\mathscr{F}_X$, $\mathscr{F}_Y$, determined by the coefficients of the respective linear representations of the two processes), but an important distinction between De Schutter-Herteleer and Melard's approach and the analysis of Section 6.3.1 is that in the former the coherency spectrum turns out to be time-dependent.

Battaglia (1979) notes that the class of oscillatory processes is not closed with respect to the sum of independent processes, and that the coherency of a bivariate oscillatory process is independent of time. In the above paper, he tries to remove these restrictions and introduces the concept of "sigma-oscillatory" processes. Within the class of such processes evolutionary spectral density functions are defined and it is proved that the resulting coherency function is then time-dependent. Moreover, this class is closed with respect to the sum of independent processes. A test for "coherency-stationarity" of bivariate sigma-oscillatory process is also derived.

6.5 SOME APPLICATIONS OF EVOLUTIONARY SPECTRAL ANALYSIS

Evolutionary spectral analysis involves the same physical ideas as those which arise in conventional spectral analysis, i.e. it aims to determine power/frequency distributions, but with the crucial distinction that here these distributions are allowed to change their form over time. The potential range of applications of evolutionary spectral analysis is thus as vast as that of conventional spectral analysis, and it is particularly suited to those situations where one encounters stochastic processes whose statistical properties are believed to be changing over time. To illustrate the practical applications of evolutionary spectral analysis we now describe some typical areas in which this approach has been successfully applied.

Analysis of jet engine noise and random vibrations

Hammond (1968, 1973) describes an interesting application of evolutionary spectral theory to the power sepctrum analysis of jet engine noise, and discusses also the application of this theory to the study of single and multi-degree of freedom systems driven by non-stationary random excitations. The spectral patterns of the sound pressure field in the vicinity of a jet engine exhaust change according to the jet velocity V. Thus if measurements are made of the field under conditions of a varying jet velocity, $V(t)$, these will correspond to a non-stationary process with a time-dependent spectrum. Hammond (1973) took a reference jet velocity of 2000 ft/sec $= V_0$, say, and analysed the changing spectral patterns as the velocity varies between 1700 and 2000 ft/sec. Evolutionary spectra were then estimated over the frequency range 100–1000 Hz. Hammond then compared the evolutionary spectra with the stationary spectra obtained when the jet velocity was held fixed at value V_j, say, V_j running from 1700 to 2000 ft/sec in steps of 50 ft/sec. By matching the resulting seven stationary spectra with the corresponding evolutionary spectra, Hammond was able to determine the form of the modulating function $A_t(\omega)$. Denoting the evolutionary spectrum at velocity $V(t)$ by $h_V(\omega)$, he writes

$$h_V(\omega) = |A_t(\omega)|^2 h_{V_0}(\omega)$$

(so that the instant at which the engine is running at velocity V_0 corresponds to "time" zero), and with this normalization the function $A_t(\omega)$ is then determined as

$$A_t(\omega) = \left\{1 - \frac{v(t)}{V_0}\right\}^4 10^{v(t)/4000} \left\{\frac{720}{900 - [\{V_0 - v(t)\}^2/22222.2]}\right\}$$
$$\times 10^{\{-(f-f_c)v(t)\}/(5300\times 10^3)}$$

Here, $v(t) = V_0 - V(t)$, $f = \omega/2\pi$ is the frequency in Hertz, and f_c is the constant frequency 365 Hz. (The value of f_c depends on the position of the point at which sound pressure measurements are made, relative to the jet exhaust.)

A sample result is shown in Fig. 6.2 below, together with the curve obtained under stationary operating conditions. The jet noise spectra are

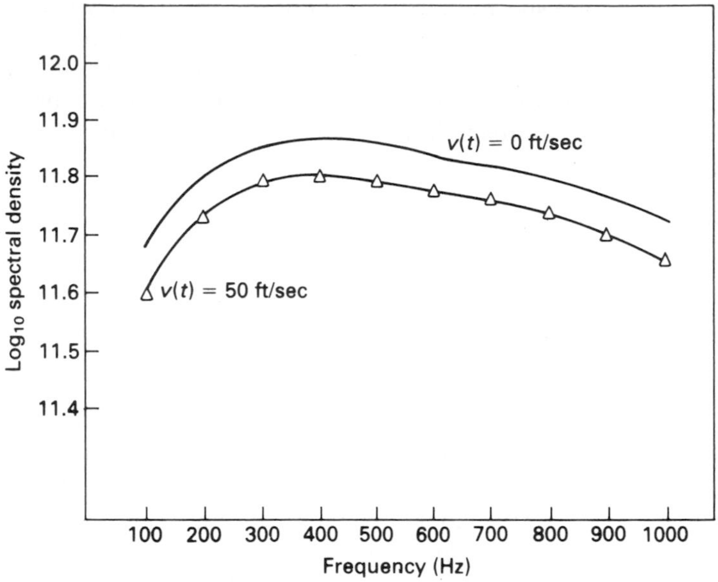

Fig. 6.2. Spectra of jet engine noise.

for $v(t)=0$ (corresponding to velocity V_0) and for $v(t)=50$ ft/sec. The continuous curves denote the "stationary" spectra, and the triangles show the values obtained by multiplying the curve for $v(t)=0$ by the function $|A_t(\omega)|^2$, with $v(t)=50$ ft/sec. The agreement is extremely good, and it may be noted that the non-stationary second pressure field process is not uniformly modulated: lower frequencies are augmented whilst higher frequencies are attenuated for lower jet velocities.

Analysis of seismological data using uniformly modulated processes

Dargahi-Noubary and Laycock (1981) discuss the problem of analysing seismological data with a view to discriminating between earthquakes and nuclear explosions. There are various types of seismological data, depending on the nature of the recorded waves. The main distinction is between "body waves" or "surface waves", and the former are further divided into "compressional" (P-waves), "shear" (S-waves), plus "reflections" (pP-waves). The surface waves are divided into "horizontal" (Love waves) and "vertical" (Rayleigh waves) types. They are also independently divided into "short period" (roughly, greater than 0.5 Hz) and "long period" (less than 0.5 Hz) records on the basis of their spectral distributions. These records have differing spectral characteristics for earthquakes and explosions; for example, there tends to be a preponderance of high-frequency energy over low-frequency energy for explosions as compared with earthquakes.

Initially, Dargahi-Noubary and Laycock model their data as zero-mean stationary Gaussian processes with two alternative spectral shapes (depending on the source of the data), and they show that the likelihood ratio approach leads to a discriminant function effectively based on the ratio of the spectral contents in two frequency bands—the so-called "spectral ratio discriminant rule". However, they subsequently observe that many seismological series exhibit clear non-stationary behaviour, and suggest that such series may be more realistically modelled as uniformly modulated processes of the form,

$$X(t) = c(t)\,Y(t)$$

where $c(t)$ is a deterministic function and $Y(t)$ is a stationary process. They show that their discrimination procedure based on spectral ratios remains valid for the above class of non-stationary processes, and particular forms of $c(t)$ which they consider are:

$$c(t) = a_1\, e^{a_2 t} + a_3\, e^{a_4 t}$$

and

$$c(t) = (a + bt)\, e^{ct}.$$

These forms have been applied to earthquake data by Fujita and Shibara (1978) with $Y(t)$ as a white noise process. Dargahi-Noubary (1978) has fitted a modulating function of the form

$$c(t) = t^{\alpha} a^{-\beta t},$$

with $Y(t)$ taken as an AR(1) process, to both earthquake and explosion data. Simulations of an earthquake and an explosion using Dargahi-Noubary's models are shown in Figs 6.3 and 6.4. A comparison of these figures with published records of real data (see, for example, Hitchings and Beresford, 1978) shows that these simulations bear a striking resemblance to the real data.

A cusum test for changes in structure of time series

Subba Rao (1981a) made use of the theory of evolutionary spectra in devising a cusum test for detecting changes in linear time series models. Suppose we are given a process $\{X_t\}$ which admits a time-dependent, one-sided, moving average representation of the form,

$$X_t = \sum_{u=0}^{\infty} b_t(u)\varepsilon_{t-u} \tag{6.5.1}$$

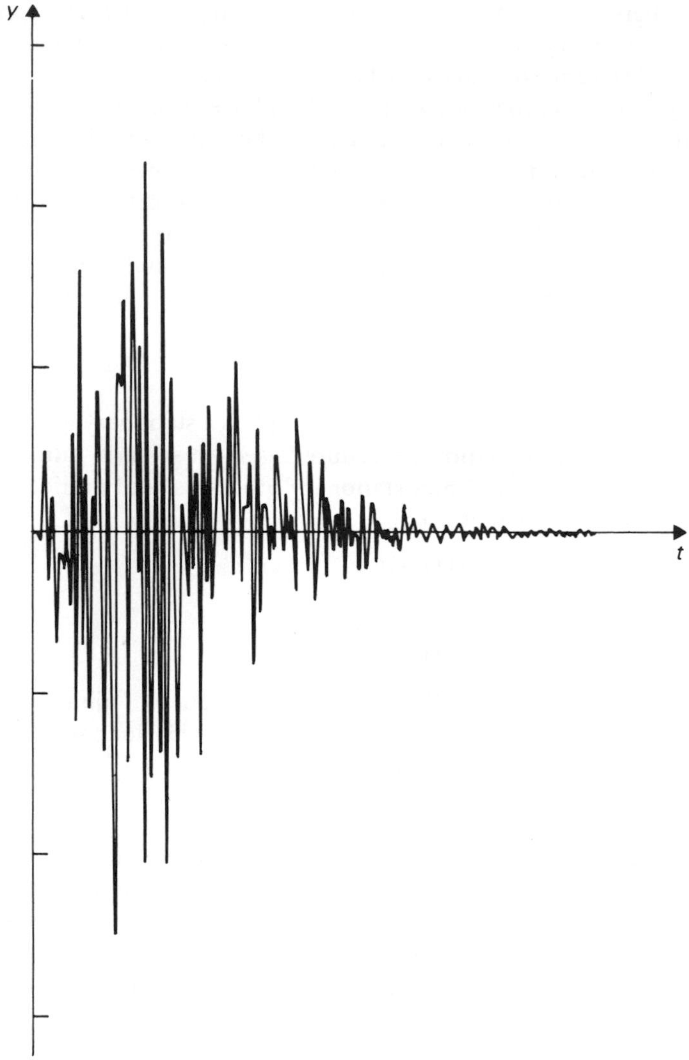

Fig. 6.3. Simulation of an earthquake as a uniformly modulated process.

where the $\{\varepsilon_t\}$ are independent normal variables with variance σ_ε^2. Under suitable conditions on the time-dependent coefficients $\{b_t(u)\}$, the evolutionary spectral density function (with respect to the family $\{\sum_{u=0}^{\infty} b_t(u)\, e^{-iu\omega}\}$) is given by

$$h_t(\omega) = \frac{\sigma_\varepsilon^2}{2\pi} \left| \sum_{u=0}^{\infty} b_t(u)\, e^{-iu\omega} \right|^2. \tag{6.5.2}$$

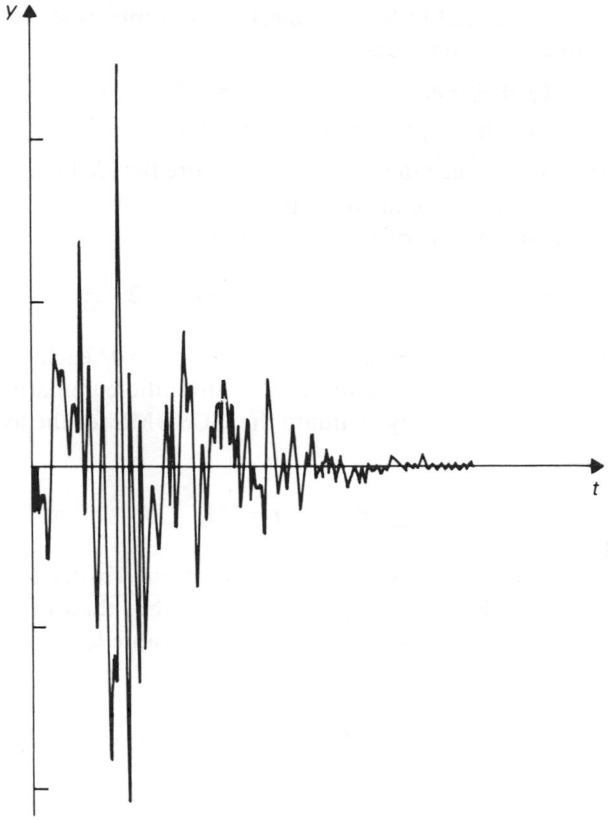

Fig. 6.4. Simulation of an explosion as a uniformly modulated process.

If the coefficients $\{b_t(u)\}$ remain constant up to time t_0, and then change abruptly after time t_0, this will be reflected in a corresponding change in the evolutionary spectral density functions. Given observations on $\{X_t\}$, we may estimate $h_t(\omega)$ over a suitable grid of frequencies $\{\omega_t\}$ and over a suitable grid of time points $\{t_i\}$. Provided that the time points $\{t_i\}$ and frequency points $\{\omega_j\}$ are each spaced sufficiently wide apart, the estimated spectra, $\hat{h}_{t_i}(\omega_j)$, then conform to a log-linear model. Specifically, writing $Y_{ij} = \log_e\{\hat{h}_{t_i}(\omega_j)\}$, we may write

$$Y_{ij} = \mu + \alpha_i + \beta_j + \gamma_{ij} + e_{ij}, \qquad i = 1, \ldots, I, j = 1, \ldots, J, \text{ say}, \quad (6.5.3)$$

where the $\{e_{ij}\}$ are approximately independent $N(0, \sigma_e^2)$ variables. (This log-linear model for the $\hat{h}_{t_i}(\omega_j)$ was first introduced by Priestley and Subba Rao (1969) in connection with their test for stationarity, and its derivation is discussed in more detail in Section 6.6.) If the series is stationary up to time t_0 then the above model becomes,

$$Y_{ij} = \begin{cases} \mu + \beta_j + e_{ij}, & i = 1, 2, \ldots, t_0 \\ \mu + \alpha_i + \beta_j + \gamma_{ij} + e_{ij}, & i = t_0 + 1, t_0 + 2, \ldots. \end{cases} \tag{6.5.4}$$

Thus, testing for a change in the model structure for $\{X_t\}$ may be effected by testing for a change in the mean of the sequence $\{Y_{1j}, Y_{2j}, \ldots, Y_{Ij}\}$, for each j. This suggests the use of the cumulative sums,

$$S_m(j) = \sum_{i=1}^{m} (Y_{ij} - \mu - \beta_j), \qquad m = 1, 2, \ldots. \tag{6.5.5}$$

If there is no change in structure then $E[S_m(j)] = 0$, each j; otherwise $E[S_m(j)] \neq 0$. Alternatively, instead of evaluating the cumulative sums for each frequency index j, we may evaluate the CUSUMS for the average, $\bar{Y}_{i.}$, given by,

$$\bar{Y}_{i.} = \frac{1}{J} \sum_{j=1}^{J} Y_{ij}, \qquad i = 1, 2, \ldots, I. \tag{6.5.6}$$

This is a more suitable procedure since we are interested in detecting whether there is a change in the overall spectral form rather than in a particular frequency component. We now choose a reference value k, say, and compute the cumulative sum,

$$S_m = \sum_{i=1}^{m} (\bar{Y}_{i.} - k), \qquad m = 1, 2, \ldots. \tag{6.5.7}$$

The reference value k is chosen *a priori*, and would usually be based on the values of spectral ordinates computed from previous series. A change of slope of the CUSUM path $(m, S_m, m = 1, 2, \ldots)$ would suggest a change in the structure of the series. For example, if there is no change then the mean value of S_m is approximately zero; if there is an increase in the mean of $\bar{Y}_{i.}$ relative to k the mean of S_m increases with m, whereas a decrease in the mean of $\bar{Y}_{i.}$ will cause a decrease in the mean of S_m as a function of m. Although such increases or decreases in the slope of the CUSUM path may be apparent from visual inspection, it is common practice to use a "V-mask" to detect these changes (see, for example, Woodward and Goldsmith, 1964). The V-mask is superimposed on the CUSUM chart, and changes in structure are then indicated by the path crossing either the upper or the lower limb of the mask. If the complete path lies within the V-mask then no change has been detected.

Subba Rao (1981a) applied the above test to the following two series.

Series A. A stationary AR(2) process generated from the model

$$Y_t - 0.8\,Y_{t-1} + 0.4\,Y_{t-2} = \varepsilon_t, \qquad t = 1, \ldots, 2000 \tag{6.5.8}$$

the $\{\varepsilon_t\}$ being independent zero mean normal variables with $\sigma_e^2 = 10^4$.
Series B. A uniformly modulated process generated by

$$X_t = c(t)\, Y_t,$$

where Y_t is the stationary AR(2) process given in series A, and $c(t)$ is given by

$$c(t) = \begin{cases} 1 & t = 1, 2, \ldots, 1000, \\ \exp\left\{-\dfrac{1}{2(200)^2}(t-500)^2\right\}, & t = 1001, \ldots, 2000. \end{cases} \tag{6.5.9}$$

The spectral density function of Y_t is

$$h_y(\omega) = \frac{\sigma_e^2}{2\pi|1 - 0.8\, e^{-i\omega} + 0.4\, e^{-2i\omega}|^2}, \tag{6.5.10}$$

while the evolutionary spectral density functions for X_t are

$$h_{X,t}(\omega) = \begin{cases} h_y(\omega), & t \le 1000, \\ |C(t)|^2 h_y(\omega), & t > 1000, \end{cases} \tag{6.5.11}$$

or equivalently,

$$\log_e h_{X,t}(\omega) = \begin{cases} \log_e h_y(\omega), & t \le 1000, \\ 2\log_e|C(t)| + \log_e h_y(\omega), & t > 1000. \end{cases} \tag{6.5.12}$$

Sample evolutionary spectra for both Y_t and X_t were computed from 2000 observations generated from each model. The estimates were computed using the "double window technique" described in Section 6.3 (cf. equation (6.3.22a)), with the sequences $\{g_u\}$ and $\{w_v\}$ chosen as

$$g_u = \begin{cases} 1/2\sqrt{h\pi}, & |u| \le h, \\ 0, & |u| > h, \end{cases} \tag{6.5.13}$$

and

$$w_v = \begin{cases} 1/T', & |v| \le T'/2, \\ 0, & |v| > T'/2, \end{cases} \tag{6.5.14}$$

with $h = 7$, $T' = 100$. The estimated spectra were then evaluated at time points $t_i = 100i$ $(i = 1, \ldots, 19)$, and frequencies $\omega_j = 2\pi j/15$ $(j = 1, 2, \ldots, 14)$. With the above choice of windows it may be shown that the variance of the residual e_{ij} in (6.5.3) is $\sigma_e^2 = 4h/3T'$ (see Priestley and Subba Rao, 1969).

Following the above discussion, the CUSUM path for series A should lie completely within the V-mask, with the values of S_m fluctuating about zero for all m. For series B, the CUSUM path should behave as for series A up to $t = 1000$, and beyond this point S_m should decrease with m, and thus cross one of the limbs of the mask. Here, the reference value k is taken as the average (over both time and frequency) of the logarithm of the evolutionary spectral estimates computed from series A, giving $k = 8.10$. In Table 6.1 we show the values of t_i, $\bar{Y}_{i.}$ $(i = 1, \dots, 19)$, and S_m $(m = 1, 2, \dots, 19)$, for series A and series B.

With the values of h, T', given above we find $\sigma_e = 0.31$. The scale for plotting the values of S_m against m were therefore chosen so that 1 unit on the horizontal scale is equal to 0.6 units on the vertical scale. It is apparent from Table 6.1 that for series A the values of S_m all fluctuate around zero, confirming that there is no change in the structure of this series. On the other hand, for series B the values of S_m show a sudden drop between t_{10} and t_{11} (corresponding to the time points 1000 and 1100). In fact, a plot of S_m intersects the V-mask, confirming that a change of structure occurred at $t = 1000$.

Upcrossings of non-stationary processes

Let $\{X(t)\}$ be a continuous parameter stationary process whose sample functions are almost surely continuously differentiable. The process is said

Table 6.1

	Series A			Series B		
	$\bar{Y}_{i.}$	$\bar{Y}_{i.} - k$	S_m	$\bar{Y}_{i.}$	$\bar{Y}_{i.} - k$	S_m
t_1	8.35	0.25	0.25	8.36	0.26	0.26
t_2	8.19	0.09	0.34	8.19	0.09	0.35
t_3	8.12	0.02	0.36	8.12	0.02	0.37
t_4	8.02	−0.08	0.28	8.03	−0.07	0.30
t_5	8.04	−0.06	0.22	8.04	−0.06	0.24
t_6	8.12	0.02	0.24	8.13	0.03	0.27
t_7	7.87	−0.23	0.01	7.87	−0.23	0.04
t_8	7.85	−0.25	−0.24	7.85	−0.25	−0.21
t_9	8.02	−0.08	−0.32	8.03	−0.07	−0.28
t_{10}	8.29	0.19	−0.13	7.66	−0.24	−0.52
t_{11}	8.25	0.15	0.02	0.88	−7.22	−7.74
t_{12}	8.02	−0.08	−0.06	−2.06	−10.16	−17.90
t_{13}	7.95	−0.15	−0.21	−5.39	−13.49	−31.39
t_{14}	8.16	0.06	−0.15	−8.93	−17.03	−48.42
t_{15}	8.05	−0.05	−0.20	−12.77	−20.87	−69.29
t_{16}	8.13	0.03	−0.17	−17.13	−25.23	−94.52
t_{17}	8.18	0.08	−0.09	−22.29	−30.39	−124.91
t_{18}	8.21	0.11	0.02	−27.09	−35.29	−160.20
t_{19}	8.16	0.06	0.08	−39.91	−48.01	−208.21

to have an "upcrossing" at level u at time t if $X(t)=u$, and $\dot{X}(t)>0$. Let $p(t; x, y)$ denote the joint probability density function of $\{X(t), \dot{X}(t)\}$. Under general conditions on $\{X(t)\}$, the expected rate of upcrossings at level u, $q_u(t)$, is given by

$$q_u(t)=\int_0^\infty yp(t; u, y)\, dy. \tag{6.5.15}$$

When $\{X(t)\}$ is a stationary process with integrated spectrum $H(\lambda)$, q_u is constant and is given by

$$q_u=\frac{1}{2\pi}\sqrt{\left(\frac{\lambda_2}{\lambda_0}\right)}\exp\left(-\frac{1}{2}\frac{u^2}{\lambda_0}\right), \tag{6.5.16}$$

where

$$\lambda_i=\int_{-\infty}^{\infty}|\lambda|^i\, dH(\lambda), \qquad i=0, 1, 2, \ldots$$

(see, for example, Cramer and Leadbetter, 1967).

The above result for the expected rate of upcrossings can be used to derive bounds for the distribution of the maximum of $X(t)$ over a finite interval $(0, T)$. This bound is reasonably accurate when $X(t)$ has a wide spectrum, but is too large when $X(t)$ has a narrow band spectrum. In the latter case a better approximation to the distribution of the maximum is obtained by considering the *envelope* of $X(t)$ in place of $X(t)$. For a zero-mean stationary Gaussian process the envelope is defined by considering first the real form of the spectral representation, namely

$$X(t)=\int_0^\infty \{\cos\lambda t\, dU(\lambda)+\sin\lambda t\, dV(\lambda)\} \tag{6.5.17}$$

(see *Spec. Anal.*, p. 251), and then defining the *quadrature process*, $\hat{X}(t)$, by

$$\hat{X}(t)=\int_0^\infty \{\sin\lambda t\, dU(\lambda)-\cos\lambda t\, dV(\lambda)\}. \tag{6.5.18}$$

The envelope, $R(t)$, is then defined by

$$R(t)=[X^2(t)+\hat{X}^2(t)]^{1/2}. \tag{6.5.19}$$

It may be shown that $R(t)$ has a Rayleigh distribution, and its derivative, $\dot{R}(t)$, has a normal distribution, and is distributed independently of $R(t)$. The expected rate of upcrossings of $R(t)$ at level u is given by

$$q_u^*=\sqrt{\left(\frac{\Delta}{2\pi}\right)}u\exp\left(-\frac{1}{2}\frac{u^2}{\lambda_0}\right), \tag{6.5.20}$$

where $\Delta=(\lambda_0\lambda_2-\lambda_1^2)/\lambda_0^3$ (see Cramer and Leadbetter, 1967).

Hasofer and Petocz (1978) have extended the above result to the case of oscillatory processes. Let $Y(t)$ be an oscillatory process with evolutionary spectral representation,

$$Y(t)=\int_{-\infty}^{\infty} e^{it\lambda}A_t(\lambda)\,dZ(\lambda). \tag{6.5.21}$$

Writing

$$A_t(\lambda)=\alpha(t,\lambda)+i\beta(t,\lambda),$$

we can rewrite the evolutionary spectral representation in the real form

$$Y(t)=\int_0^{\infty}\{\cos t\lambda\,dU_t^*(\lambda)+\sin t\lambda\,dV_t^*(\lambda)\} \tag{6.5.22}$$

where

$$dV_t^*(\lambda)=\alpha(t,\lambda)\,dU(\lambda)+\beta(t,\lambda)\,dV(\lambda),$$
$$dV_t^*(\lambda)=\alpha(t,\lambda)\,dV(\lambda)-\beta(t,\lambda)\,dU(\lambda),$$

with $dU(\lambda)=2\mathscr{R}[dZ(\lambda)]$ $(\lambda\neq 0)$, $dV(\lambda)=-2\mathscr{I}[dZ(\lambda)]$.

The quadrature process, $\hat{Y}(t)$, may then be defined as above, namely

$$\hat{Y}(t)=\int_0^{\infty}\{\sin\lambda t\,dU^*(\lambda)-\cos\lambda t\,dV_t^*(\lambda)\}, \tag{6.5.23}$$

and the envelope process, $R(t)$, is then defined by

$$R(t)=[Y^2(t)+\hat{Y}^2(t)]^{1/2}. \tag{6.5.24}$$

Hasofer and Petocz (1978) derive the joint distribution of $\{R(t),\dot{R}(t)\}$, and then apply the general result (6.5.15) to obtain the expected rate of upcrossings at level u of $R(t)$. Denoting the rate of upcrossings at level u by $q_u(t)$, they show that

$$q_u(t)=\left(\frac{\Delta}{2\pi}\right)^{1/2}u\exp\left[-\frac{u^2}{2}\left(1+\frac{\tau_0^2}{\Delta}\right)\right]$$
$$+\Phi\left(\frac{\tau_0 u}{\Delta^{1/2}}\right)\tau_0 u^2\exp\left[-\frac{u^2}{2}\right], \tag{6.5.25}$$

where

$$\tau_0=\int_0^{\infty}(\alpha^2+\beta^2)\,dG(\lambda)$$
$$\tau_1=\int_0^{\infty}[\lambda(\alpha^2+\beta^2)+(\alpha\dot{\beta}-\beta\dot{\alpha})]\,dG(\lambda),$$
$$dG(\lambda)=E[\{dU(\lambda)\}^2],$$
$$\Delta=1-\tau_0^2-\tau_1^2.$$

In the stationary case, $A_t(\lambda)$ is independent of t so that $\dot{\alpha} = \dot{\beta} = 0$. It is easy to verify that (6.5.25) then reduces to (6.5.20). However, in general (6.5.25) and (6.5.20) yield quite different results for high level u. Thus, stationary and non-stationary processes behave quite differently in respect of the distributions of their upcrossings and maxima.

Adaptive smoothing using evolutionary spectra

Exponential smoothing is a widely used simple technique for forecasting stationary time series data. If $\{X_t\}$ is a stationary process, fluctuating about a constant mean level, then given observations up to time t, the forecast for X_{t+1} is given by the exponentially smoothed value at time t of the series S_t, where

$$S_t = \alpha X_t + (1-\alpha) S_{t-1}. \tag{6.5.26}$$

Here α is smoothing constant, and in practice typical values of α range between 0.1 and 0.2. Such values of α provide reasonably stable forecasts provided the process is stationary, but if there is a sudden change in structure (such as, for example, a "spike" or "step" change) then the forecasts will lag behind the given series for a substantial period. However, if a value of α close to 1 is used near the change point then the smoothed series will adapt to the given series quite rapidly. Once the smoothed series has been so adjusted, a small value of α would again be appropriate. It is thus desirable to have some procedure whereby the value of α is increased when a change in structure is detected, but is reduced to the normal range 0.1 to 0.2 otherwise.

Trigg and Leach (1967) proposed the use of a "tracking signal" to detect such changes in structure and use this signal to adjust the value of α. However, Rao and Shapiro (1970) developed a more comprehensive approach in which the value of α is continuously adjusted according to the local spectral properties of the process. More specifically, they evaluate evolutionary spectral estimates at successive time points, and base the current value of α on a weighted average of the logarithm of the spectral estimates at three local time points. Let $\{\omega_k\}$ denote a set of frequencies covering the range $(0, \pi)$, let $\hat{h}_t(\omega_k)$ denote the evolutionary spectral estimate at time t and frequency ω_k, and let $Y_{t,k} = \log_e\{\hat{h}_t(\omega_k)\}$. (The reason for using a logarithmic transformation is to stabilize the variance of $\hat{h}_t(\omega_k)$—see Section 6.6.) Now write

$$\delta_{t,k} = \tfrac{1}{3}(Y_{t-2,k} + Y_{t-1,k} + Y_{t,k}) - Y_{t,k} \tag{6.5.27}$$

and

$$\Delta_t = \max_k |\delta_{t,k}|. \quad (6.5.28)$$

Thus, Δ_t is a measure of the maximum change in the (logarithmic) smoothed spectra over three time points. If the value of Δ_t is small (compared with its standard deviation) we may infer that no change in structure has occurred, and thus a small value of α is appropriate. On the other hand, if Δ_t is large, then the value of α should be chosen near to 1. Rao and Shapiro propose the following algorithm; let

$$\beta_t = b + c(\Delta_t/\sigma)^2, \quad (6.5.29)$$

where $\sigma^2 = \text{var}(\Delta_t)$, and b, c are suitable chosen constants. The value of α at the time t, denoted by α_t, is then determined by

$$\alpha_t = \max[0.1, \min(e^{\beta_t} - 1, 1)]. \quad (6.5.30)$$

Rao and Shapiro apply their technique to various simulated series which contain both "spike" and "step" changes. They conclude that the spectral approach performs considerably better, in terms of forecasting accuracy, than the method of Trigg and Leach. The improvement is particularly marked in the case of "spike" disturbances.

Applications of evolutionary spectral analysis to oceanography

Tayfun *et al.* (1971) discuss the application of evolutionary spectral analysis to the study of oceanographic records. They point out that conventional methods of spectral analysis, based on the assumption of stationarity, would be untenable in, for example, the case of a hurricane storm when the extreme wave field changes rapidly with time. A similar approach, also based on the use of evolutionary spectral analysis, was adopted by Brant and Shinozuka (1969) in their simulation study of earthquakes.

6.6 TEST FOR STATIONARITY

Various methods have been proposed for testing whether or not a given series may be regarded as stationary. Some of these are designed to detect non-stationary "trends" in a particular characteristic of the series, such as the mean or the variance (Granger and Hatanaka, 1964; Grenander and Rosenblatt, 1957; Parthasarathy, 1961; Sen, 1965; Subba Rao, 1968), whilst others are designed to test whether the covariance or spectral properties of two sections of a series are compatible (Jenkins, 1961; Quenouille, 1958).

In the latter case the two sections of the series have to be specified, *a priori*, and it is assumed that within each section the series is stationary. The method to be described in this section differs from those referred to above in that it may be used to test the *overall* stationarity of the complete second-order properties of a series. This test was proposed by Priestley and Subba Rao (1969), and the basis of the method is to regard the series as non-stationary, estimate its evolutionary spectra over a range of time points, and then test these spectra for uniformity over time. The resulting test uses the techniques of a two-factor analysis of variance model. A further advantage of this approach is that it enables one to test not only the overall stationarity of the series, but also to examine the character of the non-stationarity when it exists.

Suppose we are given observations on a (possibly) non-stationary zero-mean continuous parameter process $\{X(t)\}$, assumed to possess evolutionary spectral density functions, $\{h_t(\omega)\}$. We may construct estimates of the $\{h_t(\omega)\}$ by the "double window" technique described in Section 6.3, using windows $\{g(u)\}$, $\{w(v)\}$ (cf (6.3.22), (6.3.22a)). Approximate expressions for the mean and variance of the resulting estimates, $\hat{h}_t(\omega)$, are given by (6.3.23a), (6.3.23b). The form of the window $\{w(v)\}$ will usually depend on some parameter T', say, so that we may denote it more precisely by $w_{T'}(v)$, and with the usual choices it turns out that typically $\int_{-\infty}^{\infty} |W_{T'}(\omega)|^2 \, d\omega = O(1/T')$, where $W_{T'}(\omega)$ is the Fourier transform of $w_{T'}(v)$, as given in the equation following (6.3.23b). Accordingly, we assume that there exists a constant C such that

$$\lim_{T'\to\infty} \left\{ T' \int_{-\infty}^{\infty} |W_{T'}(\omega)|^2 \, d\omega \right\} = C. \tag{6.6.1}$$

In this case (6.3.23b) may be replaced by

$$\operatorname{var}\{\hat{h}_t(\omega)\} \sim \frac{C\{\tilde{h}_t^2(\omega)\}}{T'} \left\{ \int_{-\infty}^{\infty} |\Gamma(\omega)|^4 \, d\omega \right\}. \tag{6.6.2}$$

Now $\hat{h}_t(\omega)$ and $\hat{h}_t^2(\omega)$ may be regarded as "smoothed" versions of $h_t(\omega)$, $h_t^2(\omega)$, respectively. If the bandwidth of $|\Gamma(\omega)|^2$ is small compared with the "frequency domain bandwidth" of $h_t(\omega)$, and if the bandwidth of $\{w_{T'}(v)\}$ is small compared with the "time domain bandwidth" of $h_t(\omega)$, then we have effectively from (6.3.23a), (6.6.2),

$$E[\hat{h}_t(\omega)] \sim h_t(\omega) \tag{6.6.3}$$

$$\operatorname{var}[\hat{h}_t(\omega)] \sim \frac{Ch_t^2(\omega)}{T'} \left\{ \int_{-\infty}^{\infty} |\Gamma(\omega)|^4 \, d\omega \right\}. \tag{6.6.4}$$

To illustrate the use of (6.6.4) suppose, for example, that we choose $\{g(u)\}$ to be of the form

$$g(u)=\begin{cases}1/\{2\sqrt{(h\pi)}\}, & |u|\le h\\ 0, & |u|>h\end{cases} \tag{6.6.5}$$

Then

$$|\Gamma(\omega)|^2=\frac{1}{\pi}\frac{\sin^2 h\omega}{h\omega^2},$$

(corresponding to the Bartlett window), and we find that

$$\int_{-\infty}^{\infty}|\Gamma(\theta)|^4\,d\theta=\frac{2}{3}\left(\frac{h}{\pi}\right)$$

Further, if we choose $w_{T'}(v)$ to be of the form

$$w_{T'}(v)=\begin{cases}1/T', & -\tfrac{1}{2}T'\le v\le\tfrac{1}{2}T',\\ 0, & \text{otherwise,}\end{cases} \tag{6.6.6}$$

(corresponding to the Daniell window), then we have

$$C=\lim_{T'\to\infty}\left\{T'\int_{-\infty}^{\infty}|W_{T'}(\omega)|^2\,d\omega\right\}=2\pi.$$

Using these results in (6.6.4) we obtain,

$$\operatorname{var}[\hat{h}_t(\omega)]\sim\frac{4h}{3T'}\,h_t^2(\omega). \tag{6.6.7}$$

The expression for the covariance between $\hat{h}_{t_2}(\omega_2)$ and $\hat{h}_{t_1}(\omega_1)$ is given in Priestley (1966). It is sufficient here to quote the result that this covariance will be effectively zero if either

(i) $|\omega_1\pm\omega_2|\gg$ bandwidth of $|\Gamma(\theta)|^2$; or
(ii) $|t_1-t_2|\gg$ "width" of the function $\{w_{T'}(v)\}$.

With the usual modifications (see Section 6.3), a similar procedure may be used to estimate the evolutionary spectra of discrete parameter processes. Note that the expression (6.6.4) for the variance of $\hat{h}_t(\omega)$ should be doubled when $\omega=0$ (and $\pm\pi$ in the discrete case).

The basis of the test

It is well known that in the case of stationary processes a logarithmic transformation will stabilize (approximately) the variance of the estimated spectral density function, and this device was suggested by Jenkins and

Priestley (1957) and Grenander and Rosenblatt (1957) in connection with goodness-of-fit tests. It follows from (6.6.3), (6.6.4) that this transformation will produce the same effect when applied to evolutionary spectral estimates. Thus, if we write

$$Y(t, \omega) = \log_e\{\hat{h}_t(\omega)\},$$

then we have approximately

$$E[Y(t, \omega)] = \log_e\{h_t(\omega)\}$$

and

$$\operatorname{var}[Y(t, \omega)] = \sigma^2, \qquad \omega \neq 0, \pm\pi,$$

where

$$\sigma^2 = C\left\{\int_{-\infty}^{\infty} |\Gamma(\omega)|^4 \, d\omega\right\} \Big/ T' \tag{6.6.8}$$

is now independent of ω and t.

Alternatively, we may write

$$Y(t, \omega) = \log\{h_t(\omega)\} + e(t, \omega),$$

where approximately

$$E[e(t, \omega)] = 0, \qquad \text{all } t, \omega,$$

$$\operatorname{var}[e(t, \omega)] = \sigma^2, \qquad \text{all } t, (\omega \neq 0, \pm\pi).$$

Suppose now that we have evaluated the estimated evolutionary spectra over the interval $(0, T)$. We choose a set of times $t_1, t_2, \ldots, t_I$ (say) and a set of frequencies $\omega_1, \omega_2, \ldots, \omega_J$ (say) which cover the range of times and frequencies of interest, and are such that conditions (i) and (ii) above are both satisfied. If now we write,

$$Y_{ij} = Y(t_i, \omega_j), \qquad h_{ij} = h_{t_i}(\omega_j), \qquad e_{ij} = e(t_i, \omega_j),$$
$$i = 1, \ldots, I; \; j = 1, \ldots, J$$

then we have the model

$$Y_{ij} = h_{ij} + e_{ij}, \tag{6.6.9}$$

and if the $\{t_i\}$ and $\{\omega_j\}$ are spaced "sufficiently wide apart" the $\{e_{ij}\}$ will be approximately uncorrelated. As yet, only the first two moments of the $\{\hat{h}_t(\omega)\}$ have been derived, but Jenkins (1961) suggested that, in the case of stationary processes, the logarithmic transformation would bring the distribution of the spectral estimates closer to normality. Although this statement was not established rigorously, its validity is rendered highly plausible by the analogous result given in Kendall and Stuart (1966, Vol. 3, p. 93) relating to the distribution of the sample variance of a set of independent normal observations. It is shown by these authors that the variance-stabilized, logarithmically transformed sample variance tends to normality more rapidly than the untransformed sample variance. The relevance of this result follows from the fact that a spectral estimate is distributed as a weighted sum of χ^2 variables, and consequently may be treated approximately as a χ^2 variable—as suggested by Blackman and Tukey (1959) and Jenkins (1961). It seems reasonable to suppose, therefore, that (at least approximately) we may treat the $\{e_{ij}\}$ as independent $N(0, \sigma^2)$. (Strictly, this result requires rather more than the approximate normality of $\log \hat{h}_t(\omega)$; it requires, in fact, that the joint distribution of $\{\hat{h}_{t_i}(\omega_j)\}$, $i=1,\ldots, I$, $j=1,\ldots, J$, be approximately normal. A rigorous proof of this would certainly require deeper study, but there seems little doubt that a result of this type could be proved under sufficiently strong conditions.)

It should be recalled that the variances of $\hat{h}_t(0)$, and $\hat{h}_t(\pi)$ (in the discrete case), are equal to $2\sigma^2$. Accordingly, these frequencies should either be omitted from the set $(\omega_1, \ldots, \omega_J)$, or alternatively for each t, $\hat{h}_t(0)$ and $\hat{h}_t(\pi)$ should be replaced by the single "entry", $\frac{1}{2}\{\hat{h}_t(0)+\hat{h}_t(\pi)\}$. This device will not affect the test for stationarity but will certainly affect the test for "randomness"—see below. With the above assumption, (6.6.9) becomes the usual *two-factor analysis of variance* model, and may be rewritten in the conventional form

$$H\colon\ Y_{ij} = \mu + \alpha_i + \beta_j + \gamma_{ij} + e_{ij}, \qquad i=1,\ldots, I;\ j=1,\ldots, J. \tag{6.6.10}$$

If $\{X(t)\}$ is a stationary process then,

$$E\{\hat{h}_t(\omega)\} \sim h(\omega) \quad \text{(independent of } t\text{)},$$

where $h(\omega)$ is the usual (non time-dependent) spectral density function. Consequently, we may test the stationarity of $\{X(t)\}$ by using standard techniques to test the model

$$H_1\colon\ Y_{ij} = \mu + \beta_j + e_{ij} \tag{6.6.11}$$

against the general model H given by (6.6.10). Note that we may test for the presence of the "interaction" term, γ_{ij}, even with one observation per "cell", since in this situation we know the value of $\sigma^2 = \text{var}\{e_{ij}\}$, *a priori.* In fact, it turns out that the interaction term has a rather interesting interpretation, and we discuss this point in the next section.

Interpretation of the parameters

The parameters $\{\alpha_i\}$, $\{\beta_j\}$ may be interpreted as the "main effects" of the time and frequency "factors" (respectively), and the $\{\gamma_{ij}\}$ represent an "interaction" between these two factors. It is interesting to enquire under what circumstances we would expect the $\{\gamma_{ij}\}$ to be all zero. If all the $\{\gamma_{ij}\}$ are in fact zero, then $\log h_t(\omega)$ is additive in terms of time and frequency, so that $h_t(\omega)$ is "multiplicative," i.e. may be written in the form,

$$h_t(\omega) = c^2(t)h(\omega), \tag{6.6.12}$$

for some functions $c(t)$, $h(\omega)$.

If $h_t(\omega)$ is of the form (6.6.12) it is not difficult to show that $\{X(t)\}$ must be of the form

$$X(t) = c(t)X_0(t) \tag{6.6.13}$$

where $\{X_0(t)\}$ is a stationary process with spectral density function, $h(\omega)$. Processes of the form (6.6.13) were discussed in Section 6.3 where they were described as *Uniformly modulated processes.* Thus *a test for the presence of interaction is equivalent to testing whether or not $\{X(t)\}$ is a uniformly modulated process.*

In addition to the test for stationarity, this approach provides also a test for *complete randomness* (i.e. constancy of spectra over frequency). This is achieved simply by testing the model

$$H_2\colon\; Y_{ij} = \mu + \alpha_i + e_{ij} \tag{6.6.14}$$

against the general model H.

The test procedure

Given the computed values of $Y_{ij} \equiv \log \hat{h}_{t_i}(\omega_j)$, we first construct the standard analysis of variance table for a two-factor design, which, with the usual notation, is set out in Table 6.2.

Table 6.2

Item	Degrees of freedom	Sum of squares
Between times	$I-1$	$S_T = J\sum_{i=1}^{I}(Y_{i.} - Y_{..})^2$
Between frequencies	$J-1$	$S_F = I\sum_{j=1}^{J}(Y_{.j} - Y_{..})^2$
Interaction + residual	$(I-1)(J-1)$	$S_{I+R} = \sum_{i=1}^{I}\sum_{j=1}^{J}(Y_{ij} - Y_{i.} - Y_{.j} + Y_{..})^2$
Total	$IJ-1$	$S_0 = \sum_{i=1}^{I}\sum_{j=1}^{J}(Y_{ij} - Y_{..})^2$

1. In testing for stationarity, the first step is to test the interaction sum of squares, using the result, $(S_{I+R}/\sigma^2)=\chi^2_{(I-1)(J-1)}$. (Recall that, since σ^2 is known, all comparisons are based on χ^2 rather than F-tests.)
2. If the interaction is not significant, we conclude that $\{X(t)\}$ is a Uniformly modulated process, and proceed to test for stationarity by testing S_T, using

$$(S_T/\sigma^2)=\chi^2_{(I-1)}.$$

3. If, however, the interaction turns out to be significant, we conclude that $\{X(t)\}$ is non-stationary, and non-uniformly modulated. As is usually the case, there is now little point in testing the "main-effect" S_T, but we may well wish to examine whether the non-stationarity of $\{X(t)\}$ is restricted only to some frequency components. (For example, we may wish to test whether the "high" frequencies are stationary and only the "low" frequencies non-stationary.) To test this type of hypothesis we select those frequencies of interest, say $\{\omega_{j_1}, \omega_{j_2}, \ldots, \omega_{j_k}\}$, and test for stationarity at these frequencies by using the statistic

$$\sum_{j\in K}\sum_{i=1}^{I}(Y_{ij}-Y_{.j})^2=\sigma^2\chi^2_{k(I-1)} \tag{6.6.15}$$

(Here, K denotes the set of integers $\{j_1, j_2, \ldots, j_k\}$.) In particular, this type of test may be used to examine whether any one particular frequency component is stationary.
4. Reversing the roles of "times" and "frequencies", the above procedure may be used in exactly the same way to test for "complete randomness", either at all times (using S_F when S_I is not significant), or at a particular subset of times (using a statistic analogous to (6.6.15)).

Examples

We now apply the test to two examples.

Example 1. Here we consider the discrete parameter uniformly modulated process,

$$X_t=\{e^{-(t-500)^2/2(200)^2}\}X_t^{(0)}, \qquad (t=0, 1, 2, \ldots)$$

where $X_t^{(0)}$ is the (stationary) second-order autoregressive process

$$X_t^{(0)}-0.8X_{t-1}^{(0)}+0.4X_{t-2}^{(0)}=\varepsilon_t$$

in which the $\{\varepsilon_t\}$ are independent $N(0, 100^2)$ variables. Realizations of this process were constructed, and the evolutionary spectra estimated for $t=108(150)558$. The estimates, $\hat{h}_t(\omega)$, were obtained by using the discrete time analogue of (6.3.22a) in which $w_{T'}(v)$ is given by (6.6.6), with $T'=200$, and $g(u)$ has the form (6.6.5) with $h=7$. For this analysis we have,

$$\sigma^2 = 7/150 \qquad \text{(cf. equation (6.6.8))}.$$

Also, the window $|\Gamma(\omega)|^2$ has a bandwidth of approximately $\pi/h=\pi/7$, and the window $\{w_{T'}(v)\}$ has width $T'=200$. Thus, in order to obtain approximately uncorrelated estimates, the points $\{\omega_j\}$, $\{t_i\}$ should be chosen so that the spacings between the $\{\omega_j\}$ are at least $\pi/7$ and the spacings between the $\{t_i\}$ are at least 200. In fact, as will be shown below, the test works quite satisfactorily even with a (uniform) $\{t_i\}$ spacing as low as 150. (This smaller spacing was used as it allowed us to include an extra value of t in Tables 6.3 and 6.5)). The $\{\omega_j\}$ were chosen as follows:

$$\omega_j = \pi j/20, \qquad j=1(3)19,$$

corresponding to a uniform spacing of $3\pi/20$ (which just exceeds $\pi/7$). The values of $\log \hat{h}_t(\omega)$ are shown in Table 6.3.

The analysis of variance is shown in Table 6.4.

As expected, the interaction is extremely small (confirming the uniformly modulated model), and both the "between times" and "between frequencies" sums of squares are highly significant, confirming that the process is non-stationary and that the spectra are non-uniform.

Table 6.3

	ω						
t	$\pi/20$	$4\pi/20$	$7\pi/20$	$10\pi/20$	$13\pi/20$	$16\pi/20$	$19\pi/20$
108	2.3597	2.3245	2.1499	1.9856	1.6258	1.4274	1.2494
258	3.1849	3.2967	3.3749	2.8425	2.3800	2.0380	2.0579
408	3.7692	3.8002	3.6135	3.1199	2.8137	2.5727	2.4673
558	3.7253	3.6672	3.5288	3.1247	2.7545	2.7050	2.4871

Table 6.4

Item	Degrees of freedom	Sum of squares	χ^2 (= sum of squares/σ^2)
Between times	3	7.6353	163.61
Between frequencies	6	6.4716	138.68
Interaction + residual	18	0.1848	3.96
Total	27	14.2917	306.25

Example 2. Here we consider a non-stationary, non-uniformly modulated, process which was generated by taking the same stationary process of Example 1 and passing it through each of the three approximately "band-pass" filters with ranges $(0, \pi/3)$, $(\pi/3, 2\pi/3)$, $(2\pi/3, \pi)$, respectively. The outputs of the filters were then multiplied by three different functions of time and recombined to form the process $\{X_t\}$.

The evolutionary spectra of $\{X_t\}$ are:

$$h_t(\omega)=\begin{cases}|C_1(t)|^2 h(\omega), & 0\le\omega\le\pi/3,\\ |C_2(t)|^2 h(\omega), & \pi/3<\omega\le 2\pi/3,\\ |C_3(t)|^2 h(\omega), & 2\pi/3<\omega\le\pi,\end{cases}$$

where

$$C_1(t)=e^{-(t-100)^2/2(200)^2},$$
$$C_2(t)=[1+(t-300)^2/275]^{-1}$$
$$C_3(t)=\frac{\pi}{2(300)^2}\,t^2\,e^{-t/300}.$$

The spectra were estimated from realizations using the same formulae and parameters as in Example 1 (so that again $\sigma^2=7/150$), and evaluated at the same values of t and ω. The values of $\log \hat{h}_t(\omega)$ are shown in Table 6.5.

The analysis of variance is shown in Table 6.6.

Table 6.5

	ω						
t	$\pi/20$	$4\pi/20$	$7\pi/20$	$10\pi/20$	$13\pi/20$	$16\pi/20$	$19\pi/20$
108	4.5786	4.6293	5.0356	4.6723	4.0922	3.8433	3.7419
258	4.1970	4.1694	4.9950	4.4738	4.2272	4.1818	4.0763
408	3.1430	3.3359	4.0464	3.9416	4.0465	4.3829	4.1709
558	2.7110	2.8195	3.7248	3.7301	4.1195	4.5665	4.1467

Table 6.6

Item	Degrees of freedom	Sum of squares	χ^2 (=sum of squares/σ^2)
Between times	3	2.3187	49.69
Between frequencies	6	1.9071	40.87
Interaction + residual	18	4.3055	92.26
Total	27	8.5313	182.82

In this case it will be noticed that the interaction term is highly significant (0.1%), confirming that $\{X_t\}$ is non-uniformly modulated. The "between times" and "between frequencies" sums of squares are also highly significant, but it is instructive here to decompose the total χ^2 for (between times + interaction) into its various frequency components, as previously suggested. We obtain results shown in Table 6.7.

Table 6.7

j	1	4	7	10	13	16	19	Total
χ^2	49.27	42.54	28.46	12.60	0.38	6.16	2.54	141.95

To test each component for stationarity we refer each entry to χ^2 on three degrees of freedom. Thus, on the basis of these results we might conclude that, whilst the low frequencies were certainly non-stationary, there was no evidence to suggest non-stationarity in the high frequencies (although χ^2 for $j = 16$ is reasonably large—cf. $\chi_3^2(0.05) = 7.81$). However, since the estimated spectra at different times appear quite close at the upper frequencies, this result is hardly surprising.

Chapter 7

Prediction, Filtering, and Control of Non-stationary Processes

7.1 REVIEW OF PREDICTION THEORY FOR STATIONARY PROCESSES

One of the most important areas of time series analysis is that which deals with the problem of "predicting" (or "forecasting") a future value of a series, given a set of observations on its past. This is a celebrated problem and the earliest studies may be traced back to the pioneering work of Kolmogorov (1941) and Wiener (1949). We consider first a fairly general formulation of the problem: we are given observations $X_t, X_{t-1}, \ldots, X_{t-n}$, on a discrete parameter process and wish to predict the value of X_{t+m} $(m > 0)$. The predictor, $\tilde{X}_{t+m}$, will obviously be based on the past observations so that we may write

$$\tilde{X}_{t+m} = \theta(X_t, X_{t-1}, \ldots, X_{t-n}),$$

for some function θ. The problem is to choose θ so that $\tilde{X}_{t+m}$ is, in some sense, "closest" to X_{t+m}. If we adopt the usual measure of closeness, namely the mean-square error,

$$\mathcal{M}(m) = E[\{X_{t+m} - \tilde{X}_{t+m}\}^2], \tag{7.1.1}$$

then the problem is to choose θ so as to minimize $\mathcal{M}(m)$. Posed in this form the problem admits an immediate solution. For we know that if we have two random variables, X, Y, and wish to predict Y by some function of X, $u(X)$, say, then the form of $u(X)$ which minimizes $E[\{Y - u(X)\}^2]$ is given by

$$u(X) = E[Y|X],$$

i.e. the best predictor of Y is the conditional expectation of Y, given X (see, for example, *Spec. Anal.*, p. 76). This result is readily extended to the case considered above, and thus the predictor which minimizes $\mathcal{M}(m)$ is given by,

$$\tilde{X}_{t+m} = E[X_{t+m} | X_t, X_{t-1}, X_{t-2}, \ldots, X_{t-n}], \tag{7.1.2}$$

i.e. the optimal predictor is the *conditional expectation of* X_{t+m}, *given* X_t, $X_{t-1}, \ldots, X_{t-n}$. This is a fundamental result but it cannot be applied unless we either know the joint distribution of $(X_{t+m}, X_t, \ldots, X_{t-n})$ or have a finite parameter model for $\{X_t\}$ from which the conditional expectations can be evaluated recursively. If we have no model for $\{X_t\}$, and if we have no precise knowledge of its distributional properties, then we could argue that from the usual considerations it would be reasonable to assume that $\{X_t\}$ is a Gaussian process, in which case all conditional expectations of the form (7.1.2) are linear in $X_t, X_{t-1}, \ldots, X_{t-n}$. In this case we may write (7.1.2) more explicitly as,

$$\tilde{X}_{t+m} = a_0 X_t + a_1 X_{t-1} + \cdots + a_n X_{t-n}, \tag{7.1.3}$$

and now it remains only to find the values of the constants $a_0, a_1, \ldots, a_n$ which minimize $\mathcal{M}(m)$.

There are thus two main approaches to prediction theory, namely:

1. Given a finite parameter model (usually a linear form) we may apply the result (7.1.2) and evaluate the conditional expectations recursively from the model. This is the basic idea underlying the *Box–Jenkins* forecasting methodology (Box and Jenkins, 1970; *Spec. Anal.*, p. 762), and the more sophisticated *Kalman filtering* technique (Kalman, 1960; *Spec. Anal.*, p. 807).
2. Alternatively, we may restrict attention to *linear predictors* of the form (7.1.3), and then use analytical methods to determine the coefficients $a_0, \ldots, a_n$, which minimize the mean-square error.

The latter approach was the one adopted by Kolmogorov and Wiener, and they each provided an analytic solution to the linear prediction problem in the case where n (the number of past observations) $\to \infty$, and under the assumption that $\{X_t\}$ is a *stationary process.* In this chapter we will extend the Kolmogorov–Wiener theory to the case of non-stationary processes, and will discuss the closely related problems of linear filtering and minimum variance control. As a prelude to the study of non-stationary processes we give below a brief review of the Kolmogorov–Wiener theory for the stationary case. (For a more detailed account see, e.g., *Spec. Anal.*, Chapter 10; Whittle, 1963; or Yaglom, 1962.)

The Kolmogorov approach

We consider first the case of a discrete parameter process, $\{X_t\}$, and assume that X_t is stationary, with finite variance and a purely continuous spectrum with spectral density function $h(\omega)$. We know from the discussion in Section 1.4 that if $h(\omega)$ satisfies the condition

$$\int_{-\pi}^{\pi} \log\{h(\omega)\}\, d\omega > -\infty \tag{7.1.4}$$

then $h(\omega)$ may be factorized in the form

$$h(\omega) = |G(e^{-i\omega})|^2 \tag{7.1.5}$$

where the function $G(z)$ may be chosen so that it *has no poles or zeros inside the unit circle,* $|z|<1$. We may therefore expand $G(z)$ in a one-sided power series,

$$G(z) = \sum_{u=0}^{\infty} g_u z^u, \tag{7.1.6}$$

and it then follows that X_t admits a one-sided linear representation of the form

$$X_t = \sum_{u=0}^{\infty} g_u \varepsilon_{t-u}, \tag{7.1.7}$$

where $\{\varepsilon_t\}$ is a purely random process with $\text{var}\{\varepsilon_t\} = 2\pi$. Note that $\text{var}\{X_t\} < \infty$ implies that $\sum_{u=0}^{\infty} g_u^2 < \infty$.

Suppose now that we are given a semi-infinite set of observations, X_t, $X_{t-1}, X_{t-2}, \ldots$, and wish to construct the linear least-squares predictor of X_{t+m}. The predictor, $\tilde{X}_{t+m}$, may be written as

$$\tilde{X}_{t+m} = \sum_{u=0}^{\infty} a_u X_{t-u}, \tag{7.1.8}$$

where $\{a_u\}$ is a sequence of constants to be determined. However, since each X_t is a linear function of $\varepsilon_t, \varepsilon_{t-1}, \ldots$, we may equally well express $\tilde{X}_{t+m}$ in the form

$$\tilde{X}_{t+m} = \sum_{u=0}^{\infty} b_u \varepsilon_{t-u}, \tag{7.1.9}$$

where $\{b_u\}$ is another sequence of constants. The problem will be solved if we determine either the $\{a_u\}$ or the $\{b_u\}$ sequence.

Now we may write X_{t+m} in the form

$$\begin{aligned} X_{t+m} &= \sum_{u=0}^{\infty} g_u \varepsilon_{t+m-u} \\ &= \sum_{u=0}^{m-1} g_u \varepsilon_{t+m-u} + \sum_{u=m}^{\infty} g_u \varepsilon_{t+m-u}. \end{aligned} \tag{7.1.10}$$

The first term represents that part of X_{t+m} which involves future ε_t's, and so represents the "unpredictable" part of X_{t+m}. The second term involves only present and past values of ε_t, and thus represents the "predictable" part of X_{t+m}. It is thus intuitively clear that the linear least-squares predictor of X_{t+m} is given by the second term in (7.1.10), i.e.

$$\tilde{X}_{t+m} = \sum_{u=m}^{\infty} g_u \varepsilon_{t+m-u} = \sum_{u=0}^{\infty} g_{u+m} \varepsilon_{t-u}. \qquad (7.1.11)$$

A formal proof of this result is eaily constructed by writing from (7.1.9), (7.1.10),

$$\begin{aligned}\mathcal{M}(m) &= E[\{X_{t+m} - \tilde{X}_{t+m}\}^2] \\ &= E\left[\left\{\sum_{u=0}^{m-1} g_u \varepsilon_{t+m-u} + \sum_{u=0}^{\infty} (g_{u+m} - b_u)\varepsilon_{t-u}\right\}^2\right] \\ &= \sigma_\varepsilon^2 \left[\left(\sum_{u=0}^{m-1} g_u^2\right) + \sum_{u=0}^{\infty} (g_{u+m} - b_u)^2\right].\end{aligned}$$

The first term is independent of the choice of $\{b_u\}$, while the second term is clearly minimized by choosing $b_u = g_{u+m} (u = 0, 1, 2, \ldots)$. With this choice of $\{b_u\}$ the second term above vanishes and we then have,

$$\sigma_m^2 = E[\{X_{t+m} - \tilde{X}_{t+m}\}^2] = \sigma_\varepsilon^2 \left(\sum_{u=0}^{m-1} g_u^2\right). \qquad (7.1.12)$$

The above expansion is called the *m-step prediction variance.* In particular, for $m = 1$,

$$\sigma_1^2 = E[\{X_{t+1} - \tilde{X}_{t+1}\}^2] = \sigma_\varepsilon^2 g_0^2,$$

and it may be shown further (*Spec. Anal.*, p. 741) that

$$\sigma_1^2 = 2\pi \exp\left\{\frac{1}{2\pi}\int_{-\pi}^{\pi} \log h(\omega)\, d\omega\right\}. \qquad (7.1.13)$$

This is a basic result which expresses the one-step prediction variance in terms of the spectral density function.

To write $\tilde{X}_{t+m}$ explicitly as a function of $X_t, X_{t-1}, \ldots$, we first write (7.1.11) in the form

$$\tilde{X}_{t+m} = G^{(m)}(B)\varepsilon_t,$$

where

$$G^{(m)}(z) = \sum_{u=0}^{\infty} g_{u+m} z^u.$$

We also have from (7.1.7),

$$\varepsilon_t = G^{-1}(B)X_t,$$

and thus we may write

$$\tilde{X}_{t+m} = A(B)X_t,$$

where

$$A(z) = G^{(m)}(z)G^{-1}(z). \tag{7.1.14}$$

If we now expand $A(z)$ as a power series in z we will obtain $\tilde{X}_{t+m}$ in the required form. Note that since $\sum_{u=0}^{\infty} g_u^2 < \infty$, $G^{(m)}(z)$ is analytic in $|z|<1$ and hence, since $G(z)$ has no zeros in $|z|<1$, $A(z)$ is analytic in $|z|<1$.

The expression for $A(z)$ can be written in an alternative form using a convenient notation introduced by Whittle (1963). For any function $K(z)$ with Laurent expansion

$$K(z) = \sum_{u=-\infty}^{\infty} k_u z^u$$

we define

$$[K(z)]_+ = \sum_{u=0}^{\infty} k_u z^u \tag{7.1.15}$$

and

$$[K(z)]_- = \sum_{u=-\infty}^{-1} k_u z^u \tag{7.1.16}$$

We may then write $G^{(m)}(z)$ as

$$G^{(m)}(z) = [z^{-m}G(z)]_+,$$

and $A(z)$ can now be written as

$$A(z) = [z^{-m}G(z)]_+ / G(z). \tag{7.1.17}$$

The function $A(z)$ is called the *predictor generating function* and $A(e^{-i\omega})$ is called the *predictor transfer function.*

Continuous-parameter process

Let $\{X(t)\}$ be a continuous-parameter stationary process whose spectral density function $h(\omega)$ exists for all ω and satisfies the *Paley-Wiener condition,*

$$\int_{-\infty}^{\infty} \frac{\log\{h(\omega)\}}{1+\omega^2}\, d\omega > -\infty. \tag{7.1.18}$$

It may then be shown that $h(\omega)$ can be factorized in the form

$$h(\omega) = |G(i\omega)|^2 \tag{7.1.19}$$

where $G(z)$ has a one-sided Laplace transform,

$$G(z) = \int_0^\infty g(u)\, e^{-uz}\, du \tag{7.1.20}$$

with $\int_0^\infty g^2(u)\, du < \infty$, and $G(z)$ is analytic and has no zeros in the right half-plane (so that $G(i\omega)$ is analytic and has no zeros in the lower half-plane). It then follows from (7.1.19) that $X(t)$ admits the linear representation

$$X(t) = \int_0^\infty g(u)\varepsilon(t-u)\, du \tag{7.1.21}$$

where $\{\varepsilon(t)\}$ is a continuous-parameter white noise process with spectral density function given (formally) by $h_\varepsilon(\omega) = 1$, all ω. Writing $dW(t) = \varepsilon(t)dt$, the process $\{W(t)\}$ has the same second-order properties as a *Wiener process* with increase of variance per unit time given by $\sigma_w^2 = 2\pi$, i.e.

$$E[\{dW(t)\}^2] = 2\pi\, dt \tag{7.1.22}$$

(see, for example, *Spec. Anal.*, p. 736).

Writing $\varepsilon(t-u) = e^{-uD}\varepsilon(t)$ (D denoting the differential operator, d/dt), (7.1.21) can be written as

$$X(t) = G(D)\varepsilon(t), \tag{7.1.23}$$

where $G(z)$ is the function given by (7.1.20).

From (7.1.21) we may write

$$\begin{aligned} X(t+m) &= \int_0^\infty g(u)\varepsilon(t+m-u)\, du \\ &= \int_0^m g(u)\varepsilon(t+m-u)\, du + \int_m^\infty g(u)\varepsilon(t+m-u)\, du. \end{aligned}$$

Given $\{X(s); s \le t\}$, we may argue, as in the discrete case, that the first term above represents the "unpredictable" part of $X(t+m)$, and it follows that the linear least-squares predictor of $X(t+m)$ is given by the second term, i.e.

$$\tilde{X}(t+m) = \int_m^\infty g(u)\varepsilon(t+m-u)\, du = \int_0^\infty g(u+m)\varepsilon(t-u)\, du. \tag{7.1.24}$$

The prediction error is

$$\{X(t+m)-\tilde{X}(t+m)\}=\int_0^m g(u)\varepsilon(t+m-u)\,du,$$

and the m-step prediction variance is

$$\sigma_m^2=E[\{X(t+m)-\tilde{X}(t+m)\}^2]=\sigma_w^2\left\{\int_0^m g^2(u)\,du\right\}. \tag{7.1.25}$$

Inverting (7.1.23) gives

$$\varepsilon(t)=G^{-1}(D)X(t), \tag{7.1.26}$$

and writing $\varepsilon(t-u)=e^{-uD}\varepsilon(t)$ in (7.1.24) and substituting (7.1.26) gives

$$\tilde{X}(t+m)=\left\{\int_0^\infty g(u+m)\,e^{-uD}G^{-1}(D)\,du\right\}X(t),$$

or,

$$\tilde{X}(t+m)=A(D)X(t), \tag{7.1.27}$$

where

$$A(z)=G^{-1}(z)\left\{\int_0^\infty g(u+m)\,e^{-uz}\,du\right\}. \tag{7.1.28}$$

Now for any function $K(z)$ with two-sided Laplace transform

$$K(z)=\int_{-\infty}^\infty k(u)\,e^{-uz}\,du,$$

we write

$$[K(z)]_+=\int_0^\infty k(u)\,e^{-uz}\,du, \qquad [K(z)]_-=\int_{-\infty}^0 k(u)\,e^{-uz}\,du.$$

We then have,

$$[e^{mz}G(z)]_+=\left[\int_0^\infty g(u)\,e^{-z(u-m)}\,du\right]_+=\left[\int_{-m}^\infty g(u+m)\,e^{-uz}\,du\right]_+$$

$$=\int_0^\infty g(u+m)\,e^{-uz}\,du.$$

Hence, $A(z)$ may be written as

$$A(z)=G^{-1}(z)[e^{mz}G(z)]_+. \tag{7.1.29}$$

[Although we started from an integral representation of $\tilde{X}(t+m)$ (cf. (7.1.24)), we must interpret the operator $A(D)$ fairly freely. In particular, we must allow $A(D)$ to be, for example, a finite polynomial in D (corresponding to the case where $\tilde{X}(t+m)$ is a linear combination of the derivatives of $X(t)$), as would be the case if $X(t)$ conforms to a finite-order AR model.]

The Wiener approach

The Wiener solution is based on a more direct approach in which we write $\tilde{X}_{t+m}$ as

$$\tilde{X}_{t+m} = \sum_{u=0}^{\infty} a_u X_{t-u}, \tag{7.1.30}$$

substitute this directly into (7.1.1) and minimize the resulting expression with respect to the $\{a_u\}$. We obtain,

$$\begin{aligned} \mathcal{M}(m) &= E[\{X_{t+m} - \tilde{X}_{t+m}\}^2] \\ &= \left[R(0) - 2 \sum_{u=0}^{\infty} a_u R(m+u) + \sum_{u=0}^{\infty} \sum_{v=0}^{\infty} a_u a_u R(u-v) \right], \end{aligned}$$

where $R(s)$ denotes the autocovariance function of X_t, i.e. $R(s) = \operatorname{cov}\{X_t, X_{t+s}\}$. Differentiating $\mathcal{M}(m)$ w.r. to a_u and setting the derivative equal to zero gives,

$$\sum_{v=0}^{\infty} a_v R(u-v) = R(m+u), \qquad u = 0, 1, 2, \ldots. \tag{7.1.31}$$

This equation, whose continuous-parameter analogue is known as the *Wiener–Hopf equation*, has the same form as the "normal equations" which would result from a least-squares regression analysis of X_{t+m} on X_t, $X_{t-1}, \ldots$. However, (7.1.31) involves an infinite number of unknown parameters and thus cannot be solved directly. Although the LHS of (7.1.31) has the form of a convolution between $\{a_u\}$ and $\{R(s)\}$, the equation cannot be solved by taking Fourier transforms of each side since (7.1.31) holds only for $u = 0, 1, 2, \ldots$ and does not hold for negative u. We must therefore pursue a different approach, and we now rewrite the expression for $\mathcal{M}(m)$ in its frequency domain form. Using the spectral representation for X_t we may write

$$X_{t+m} = \int_{-\pi}^{\pi} e^{i(t+m)\omega}\, dZ(\omega), \tag{7.1.32}$$

and from (7.1.30) we also have,

$$\tilde{X}_{t+m} = \int_{-\pi}^{\pi} e^{it\omega} A(e^{-i\omega})\, dZ(\omega), \tag{7.1.33}$$

where, as before,

$$A(z) = \sum_{u=0}^{\infty} a_u z^u. \tag{7.1.34}$$

Hence,

$$X_{t+m} - \tilde{X}_{t+m} = \int_{-\pi}^{\pi} e^{it\omega}\{e^{im\omega} - A(e^{-i\omega})\}\, dZ(\omega).$$

Using the orthogonality of the $\{dZ(\omega)\}$, together with $E[|dZ(\omega)|^2] = h(\omega)d\omega$, we obtain,

$$\mathcal{M}(m) = \int_{-\pi}^{\pi} |e^{im\omega} - A(e^{-i\omega})|^2 h(\omega)\, d\omega. \tag{7.1.35}$$

The problem of finding that sequence $\{a_u\}$ which minimizes $\mathcal{M}(m)$ now becomes one of finding that function $A(e^{-i\omega})$ which minimizes (7.1.35). However, we must remember that $A(e^{-i\omega})$ cannot be chosen freely. Since $\tilde{X}_{t+m}$ has to be a function of present and past observations only, $A(e^{-i\omega})$ must be a *backward transform*, i.e. it must have *a one-sided Fourier series involving only negative powers of* $e^{i\omega}$. To determine the optimum form of $A(e^{-i\omega})$ within this class of functions we use the canonical factorization of $h(\omega)$ (as given by (7.1.15)), and substituting this in (7.1.35) gives,

$$\mathcal{M}(m) = \int_{-\pi}^{\pi} |e^{im\omega} G(e^{-i\omega}) - A(e^{-i\omega}) G(e^{-i\omega})|^2\, d\omega. \tag{7.1.36}$$

By the construction of $G(z)$, $G(e^{-i\omega})$ is a backward transform, and it is then easily seen that $A(e^{-i\omega})G(e^{-i\omega})$ is also a backward transform. It is intuitively clear at this stage that $\mathcal{M}(m)$ is minimized by choosing $A(e^{-i\omega})$ so that $A(e^{-i\omega})G(e^{-i\omega})$ is the "backward part" of $\{e^{im\omega}G(e^{-i\omega})\}$. To prove this result formally we decompose $\{e^{im\omega}G(e^{-i\omega})\}$ into the sum of its backward and forward transform, i.e. we write,

$$e^{im\omega} G(e^{-i\omega}) = G_1(e^{-i\omega}) + G_2(e^{-i\omega}) \tag{7.1.37}$$

where, using the notation of (7.1.15), (7.1.16),

$$G_1(z) = [z^{-m} G(z)]_+, \qquad G_2(z) = [z^{-m} G(z)]_-.$$

Substituting (7.1.37) in (7.1.36), and using the result that for any backward transform $a(e^{-i\omega})$ and any forward transform $b(e^{-i\omega})$,

$$\int_{-\pi}^{\pi} a(e^{-i\omega})b(e^{i\omega})\,d\omega = 0,$$

we obtain finally,

$$\mathcal{M}(m) = \int_{-\pi}^{\pi} |G_2(e^{-i\omega})|^2\,d\omega + \int_{-\pi}^{\pi} |G_1(e^{-i\omega}) - A(e^{-i\omega})G(e^{-i\omega})|^2\,d\omega. \tag{7.1.38}$$

The first term is independent of the choice of $A(e^{-i\omega})$, and the second term is obviously minimized by choosing

$$A(e^{-i\omega}) = G_1(e^{-i\omega})/G(e^{-i\omega}). \tag{7.1.39}$$

Writing $z = e^{-i\omega}$ and using the above expression for $G_1(z)$ we may write (7.1.39) as

$$A(z) = [z^{-m}G(z)]_+/G(z), \tag{7.1.40}$$

which is identical to the Kolmogorov solution (7.1.17).

We note that the Fourier coefficients of $\{e^{im\omega}G(e^{-i\omega})\}$ are given by

$$\frac{1}{2\pi}\int_{-\pi}^{\pi} e^{i\omega'(m+u)}G(e^{-i\omega'})\,d\omega', \qquad u = 0, \pm 1, \pm 2, \ldots.$$

Hence, (7.1.39) may be written explicitly as

$$A(e^{-i\omega}) = \sum_{u=0}^{\infty} e^{-i\omega u}\left\{\int_{-\pi}^{\pi} e^{i\omega'(m+u)}G(e^{-i\omega'})\,d\omega'\right\}\Big/ 2\pi G(e^{-i\omega}). \tag{7.1.41}$$

The m-step prediction variance is given by

$$\sigma_m^2 = E[\{X_{t+m} - \tilde{X}_{t+m}\}^2] = \int_{-\pi}^{\pi} |G_2(e^{-i\omega})|^2\,d\omega. \tag{7.1.42}$$

Continuous-parameter processes

In this case the predictor takes the form

$$\tilde{X}(t+m) = \int_0^{\infty} a(u)X(t-u)\,du,$$

and substituting this in (7.1.1) and minimizing the resulting expression with respect to $a(u)$ we obtain

$$\int_0^{\infty} a(v)R(u-v)\,dv = R(u+m), \qquad u \geq 0, \tag{7.1.43}$$

(cf. (7.1.31)). As in the discrete parameter case, this equation cannot be solved by standard transform techniques since (7.1.43) holds only for $u \geq 0$. However, proceeding as in the previous case we can write

$$\mathcal{M}(m) = \int_{-\infty}^{\infty} |e^{im\omega} - A(i\omega)|^2 h(\omega)\, d\omega, \tag{7.1.44}$$

where $A(z) = \int_0^\infty a(u)e^{-uz}\, du$. Using the canonical factorization of $h(\omega)$ given by (7.1.19), (7.1.44) becomes,

$$\mathcal{M}(m) = \int_{-\infty}^{\infty} |e^{im\omega}G(i\omega) - G(i\omega)A(i\omega)|^2\, d\omega. \tag{7.1.45}$$

If we now decompose $\{e^{im\omega}G(i\omega)\}$ into the sum of its backward and forward transforms, i.e. if we write

$$e^{im\omega}G(i\omega) = [e^{im\omega}G(i\omega)]_+ + [e^{im\omega}G(i\omega)]_-,$$

and substitute this into (7.1.45) we find,

$$\mathcal{M}(m) = \int_{-\infty}^{\infty} |[e^{im\omega}G(i\omega)]_-|^2\, d\omega + \int_{-\infty}^{\infty} |[e^{im\omega}G(i\omega)]_+ - A(i\omega)G(i\omega)|^2\, d\omega.$$

Clearly, $\mathcal{M}(m)$ is minimized by choosing

$$A(z) \doteq [e^{mz}G(z)]_+/G(z), \tag{7.1.46}$$

which agrees with the Kolmogorov solution (7.1.29). Corresponding to (7.1.41), $A(i\omega)$ may be written explicitly as,

$$A(i\omega) = \frac{1}{2\pi G(i\omega)} \int_0^\infty e^{-i\omega u} \left\{ \int_{-\infty}^{\infty} e^{i\omega'(u+m)} G(i\omega')\, d\omega' \right\} du. \tag{7.1.47}$$

Illustrative examples of the application of both the Kolmogorov and Wiener predictors are given in *Spec. Anal.*, pp. 742–54.

7.2 THE PREDICTION OF NON-STATIONARY PROCESSES

The prediction problem for non-stationary processes has received relatively little attention. There have been a few isolated attempts to deal with this topic, but in the main the approaches have been either too general or too restricted to be useful in practical applications. For example, Parzen (1961) solved the non-stationary prediction problem in principle but his approach is an abstract one, and his solution for the optimum predictor is expressed as a certain inner-product in a Hilbert space. Cramer (1961a, 1961b) considered the same problem, and obtained some interesting results in the form of existence theorems, but did not present a method for determining the explicit form of a predictor in terms of the observed variables. Similar remarks apply to the work of Davis (1952). On the other hand, there have appeared several papers written from an engineering standpoint (see, for example, Booton, 1952; Zadeh, 1953; Bendat, 1956), but in most cases the "general solution" stops with the construction of the well-known integral equation (cf. (7.1.43)) involving the covariance function of the process and the unknown coefficients of the optimal predictor, the solution of which is obtainable only when the process obeys some very simple model. From the point of view of practical application, it would appear that the most useful results so far obtained are due to Whittle (1965), who considered non-stationary processes generated by autoregressive models with time-dependent coefficients, and obtained explicit recursive relations for the optimal predictors. In fact, some of our results for these particular models correspond very closely to those obtained by Whittle.

The success of classical prediction theory for stationary processes is due essentially to the fact that such processes admit a spectral representation in terms of an orthogonal process. This feature not only simplifies the solution of the prediction problem, but also enables one to treat a general class of stationary processes by means of a "canonical" representation, so that the discussion need not be restricted to particular models, such as the autoregressive, moving-average, and so on. It turns out that the theory of evolutionary spectra provides an ideal framework for the formulation and solution of non-stationary prediction problems. In fact, by using evolutionary spectral representations one obtains a prediction theory which is almost an exact parallel of the Wiener–Kolmogorov theory. The basic idea underlying this approach is the introduction of the evolutionary spectrum of a non-stationary process, whose form completely determines the values or the "coefficients" of the optimal linear predictor. This means that even if we are presented with observations from a process whose structure is completely unknown we may still usefully apply this prediction theory by first estimating the evolutionary spectrum. Of course, this estimation procedure introduces further complications which will not be discussed here. In this section we assume throughout that the second-order properties of the process are known *a priori*. The approach which we present below was introduced by Abdrabbo and Priestley (1967).

Linear representation for oscillatory processes

As in the prediction theory of stationary processes, we begin by constructing a one-sided linear representation for a class of oscillatory processes, the distinction with the stationary case being that the coefficients in the moving-average scheme are now time-dependent.

Discrete-parameter processes

Suppose that X_t has a representation of the form (6.3.20), and that μ is absolutely continuous with respect to $d\omega$. Then we may write the evolutionary spectral density function $h_t(\omega)$ in the form

$$h_t(\omega) = |A_t(\omega)|^2 h(\omega), \tag{7.2.1}$$

where $h(\omega) = d\mu/d\omega$. Note that $h(\omega) \equiv h_0(\omega)$ must be integrable. Suppose now that

$$(\mathrm{C}_1) \qquad \int_{-\pi}^{\pi} \log h(\omega)\, d\omega > -\infty. \tag{7.2.2}$$

Then it follows that (cf. Section 7.1) there exists a function $\psi(\omega)$ such that

$$|\psi(\omega)|^2 = h(\omega), \tag{7.2.3}$$

where $\psi(\omega)$, considered as a function of the complex variable $z = e^{i\omega}$, has no poles or zeros inside the unit circle, $|z| < 1$. The function $\psi(\omega)$ may be written as a *one-sided Fourier transform*,

$$\psi(\omega) = \sum_{u=0}^{\infty} e^{-i\omega u} g^*(u), \tag{7.2.4}$$

for some sequence $g^*(u)$.

Suppose further that:

(C_2) the family can be chosen so that, for each t, $A_t(z)$ (considered as a function of z) also has no poles or zeros inside the unit circle, so that, for each t, we may write $A_t(\omega)$ in the form,

$$A_t(\omega) = \sum_{u=0}^{\infty} e^{-i\omega u} l_t(u). \tag{7.2.5}$$

(Note that $A_t(\omega)$ is square-integrable, since $|A_t(\omega)|^2 h(\omega) = h_t(\omega)$ is integrable, and by condition C_1, $h(\omega)$ may vanish on at most a set of zero measure.)

Note further that a necessary condition for the validity of (7.2.5) is

$$(\mathrm{C}_3) \qquad \int_{-\pi}^{\pi} \log|A_t(\omega)|^2\, d\omega > -\infty, \qquad \text{all } t. \tag{7.2.6}$$

Now write X_t in the form,

$$X(t) = \int_{-\pi}^{\pi} e^{it\omega} \alpha_t(\omega)\, dz(\omega), \tag{7.2.7}$$

where $\alpha_t(\omega) = A_t(\omega)\psi(\omega)$, so that,

$$|\alpha_t(\omega)|^2 = h_t(\omega), \tag{7.2.8}$$

and $z(\omega)$ is an orthogonal process with $E|dz(\omega)|^2 = d\omega$. Since both $\psi(\omega)$ and $A_t(\omega)$ have one-sided Fourier transforms, it follows that, for each t, $\alpha_t(\omega)$ has a one-sided Fourier transform, i.e. we may write

$$\alpha_t(\omega) = \sum_{u=0}^{\infty} e^{-i\omega u} g_t(u). \tag{7.2.9}$$

Note that a necessary conditon for the validity of (7.2.9) is

$$(C_4) \qquad \int_{-\pi}^{\pi} \log h_t(\omega) d\omega > -\infty, \qquad \text{all } t. \tag{7.2.10}$$

In fact, it is readily seen that $C_4 \Leftrightarrow C_1$ and C_3, but whereas in the stationary case C_4 is both necessary and sufficient to ensure the validity of (7.2.7) and (7.2.9), it does not seem possible to prove sufficiency in the non-stationary case. Conditon C_4 implies that there exist functions $\{\alpha_t^*(\omega)\}$, say, satisfying (7.2.9) and such that, for each t, $h(\omega) = |\alpha_t^*(\omega)|^2$, so that $A_t(\omega) = e^{i\lambda(t,\omega)}\alpha_t^*(\omega)$, for some function $\lambda(t, \omega)$. In the stationary case, $\exp\{i\lambda(t, \omega)\}$, being independent of t, may be incorporated with $dz(\omega)$, and sufficiency follows. In the non-stationary case it would appear that further conditions are required. (It is interesting to note, that Cramers's result (Cramer, 1961a, Theorem 6) for the prediction of harmonizable processes has the same feature. That is, Cramer has proved only that when the spectrum of a harmonizable process satisfies a condition of the type C_4, the process must be "deterministic", i.e. that in such cases on one-sided moving-average representation is *not* admissible.) Now define the process ξ_t by

$$\xi_t = \int_{-\pi}^{\pi} e^{it\omega}\, dz(\omega), \qquad t = 0, \pm 1, \pm 2, \ldots \tag{7.2.11}$$

so that $\{\xi_t\}$ is a stationary uncorrelated process, with

$$\left.\begin{aligned} E\{\xi_t\} &= 0, & &\text{all } t,\\ E\{|\xi_t|^2\} &= 2\pi, & &\text{all } t,\\ E\{\xi_t\xi_s^*\} &= 0, & &t \neq s. \end{aligned}\right\} \tag{7.2.12}$$

It now follows from (7.2.7) and (7.2.9) that X_t may be written

$$X_t = \sum_{u=0}^{\infty} g_t(u)\xi_{t-u}, \tag{7.2.13}$$

the above expression existing as a mean-square limit since,

$$2\pi \sum_{u=0}^{\infty} g_t^2(u) = \int_{-\pi}^{\pi} |\alpha_t(\omega)|^2\, d\omega = \int_{-\pi}^{\pi} h_t(\omega)\, d\omega = \text{var}\{X_t\} < \infty \tag{7.2.14}$$

We thus have

Theorem. Let $\{X_t\}$ be a discrete-parameter oscillatory process. If there exists a family $\mathscr{F}$ satisfying condition C_2, and with respect to which $\{X_t\}$ has an absolutely continuous evolutionary spectrum satisfying condition C_4, $\{X_t\}$ may be represented as a one-sided moving average process of the form (7.2.13). Conversely, if $\{X_t\}$ has a one-sided moving average representation of the form (7.2.13), condition C_4 must be satisfied.

Continuous-parameter processes

As in the case of stationary processes, the results for discrete-parameter processes can readily be adapted to the continuous case. The measure μ is again assumed to be absolutely continuous and, in place of C_1, we assume that (using the same notation as above)

$$(C_1^*) \qquad \int_{-\infty}^{\infty} \{\log h(\omega)/(1+\omega^2)\}\, d\omega > -\infty. \qquad (7.2.15)$$

Then there exists a function $\psi(\omega)$ such that $|\psi(\omega)|^2 = h(\omega)$, $\psi(z)$ having no poles or zeros in the lower half-plane. The function $\psi(\omega)$ may now be written as a one-sided Fourier integral. Corresponding to (7.2.5) we assume now that, for each t, $A_t(\omega)$ may be written in the form

$$(C_2^*) \qquad A_t(\omega) = \int_0^{\infty} e^{-i\omega u}\, l_t(u)\, du, \qquad (7.2.16)$$

a necessary condition being

$$(C_3^*) \qquad \int_{-\infty}^{\infty} \{\log |A_t(\omega)|^2/(1+\omega^2)\}\, d\omega > -\infty, \quad \text{all } t. \qquad (7.2.17)$$

It then follows that we may write $X(t)$ in the form

$$X(t) = \int_{-\infty}^{\infty} e^{it\omega} \alpha_t(\omega)\, dz(\omega), \qquad (7.2.18)$$

where $z(\omega)$ is an othogonal process, with $E|dz(\omega)|^2 = d\omega$, $\alpha_t(\omega)$ has a one-sided Fourier integral representation of the form

$$\alpha_t(\omega) = \int_0^{\infty} e^{-i\omega u} g_t(u)\, du, \qquad (7.2.19)$$

and $|\alpha_t(\omega)|^2 = h_t(\omega)$.

Corresponding to (7.2.10), a necessary condition for the validity of (7.2.19) is

$$(C_4^*) \qquad \int_{-\infty}^{\infty} \{\log h_t(\omega)/(1+\omega^2)\}\, d\omega > -\infty, \quad \text{all } t. \qquad (7.2.20)$$

We now define the process $\xi(t)$ by

$$\xi(t) = \int_{-\infty}^{\infty} e^{i\omega t}\, dz(\omega),$$

so that

$$\left.\begin{aligned} E\{\xi(t)\} &= 0, && \text{all } t, \\ E\{|d\xi(t)|^2\} &= 2\pi\, dt, && \\ E\{d\xi(t)\, d\xi^*(s)\} &= 0, && s \neq t. \end{aligned}\right\} \tag{7.2.21}$$

Then formally we may write $X(t)$ in the form

$$X(t) = \int_0^{\infty} g_t(u)\xi(t-u)\, du \tag{7.2.22}$$

or, more precisely,

$$X(t) = \int_0^{\infty} g_t(u)\, d\xi(t-u). \tag{7.2.23}$$

As in discrete-parameter case, $\text{var}\{X(t)\} < \infty$ implies

$$\int_0^{\infty} g_t^2(u)\, du < \infty, \qquad \text{all } t.$$

7.2.1 The Time-domain Approach

Discrete-parameter processes

We assume that X_t has a one-sided linear representation of the form (7.2.13), so that we may write

$$X_t = \sum_{u=0}^{\infty} g_t(u)\xi_{t-u} = \sum_{u=-\infty}^{t} g_t(t-u)\xi_u. \tag{7.2.24}$$

Then

$$X_{t+m} = \sum_{u=-\infty}^{t} g_{t+m}(t+m-u)\xi_u + \sum_{u=t+1}^{m} g_{t+m}(t+m-u)\xi_u. \tag{7.2.25}$$

Now the predictor, $\tilde{X}_{t+m}$, is to be chosen as a linear combination of $\{X_s, s \leq t\}$ of the form

$$\tilde{X}_{t+m} = \sum_{s=-\infty}^{t} b_t(s)X_s. \tag{7.2.26}$$

However, as each X_t is a linear combination of the $\{\xi_{t-u}\}$ $(u \geq 0)$, we may equally well express $\tilde{X}_{t+m}$ as a linear combination of $\{\xi_s, s \leq t\}$ of the form

$$\tilde{X}_{t+m} = \sum_{u=-\infty}^{t} a_t(u)\xi_u. \tag{7.2.27}$$

The problem now reduces to finding the values of the coefficients $\{a_t(u)\}$ which minimize $\mathcal{M}(m)$, and, exactly as in the stationary case, the solution follows immediately from (7.2.25). For, in view of the orthogonality of the ξ's it follows that

$$\tilde{X}_{t+m} = \sum_{u=-\infty}^{t} g_{t+m}(t+m-u)\xi_u = \sum_{u=m}^{\infty} g_{t+m}(u)\xi_{t+m-u}, \tag{7.2.28}$$

so that, in terms of (7.2.27),

$$a_t(u) = g_{t+m}(t+m-u), \tag{7.2.29}$$

for all u. The prediction variance, $\mathcal{M}(m)$, is given by

$$\mathcal{M}(m) = 2\pi \sum_{u=0}^{m-1} g_{t+m}^2(u). \tag{7.2.30}$$

In practice, we observe only the values of X_s $(s \leq t)$, so that in order to compute $\tilde{X}_{t+m}$ we must express (7.2.28) in terms of the Xs rather than the ξ's. Now write

$$\xi_t = \sum_{v=0}^{\infty} k_t(v)X_{t-v}. \tag{7.2.31}$$

Then, substituting (7.2.31) in (7.2.24) and equating coefficients of ξ_t, we obtain

$$\sum_{v=0}^{p} g_{t-v}(p-v)k_t(v) = \delta_{p,0}, \qquad p = 0, 1, 2, \ldots, \text{ all } t. \tag{7.2.32}$$

Thus, given the sequence $\{g_t(u)\}$, for all t, we may, in principle, solve (7.2.32) for $\{k_t(v)\}$. When $X(t)$ is stationary, $g_t(u)$ and $k_t(v)$ are both independent of t and (7.2.32) is easily solved by introducing the generating functions of the two sequences. In general, there does not appear to be any systematic relationship between corresponding terms of the two sequences, and the most practical approach would seem to be the method of repeated back-substitution. That is, if we write (7.2.32) in detail we have

$$p = 0: \quad g_t(0)k_t(0) = 1, \qquad \text{all } t,$$

$$p = 1: \quad g_t(1)k_t(0) + g_{t-1}(0)k_t(1) = 0, \qquad \text{all } t,$$

$$p = 2: \quad g_t(2)k_t(0) + g_{t-1}(1)k_t(1) + g_{t-2}(2)k_t(2) = 0, \qquad \text{all } t, \text{ etc.}$$

This system of equations is easily solved, step by step, as it is triangular. Thus we find, for all t

$$k_t(0)=1/g_t(0),$$
$$k_t(1)=-g_t(1)/g_t(0)g_{t-1}(0),$$
$$k_t(2)=\{-g_t(2)/g_t(0)g_{t-2}(2)\}+\{g_t(1)g_{t-1}(1)/g_t(0)g_{t-1}(0)g_{t-2}(0)\},\text{ etc.}$$

Alternatively, we may substitute (7.2.24) in (7.2.31) and equating coefficients, we obtain

$$\sum_{v=0}^{p} k_{t-v}(p-v)g_t(v)=\delta_{p,0} \qquad p=0,1,2,\ldots,\text{ all } t. \tag{7.2.33}$$

Whittle (1965) has pointed out that two systems of equations of the form (7.2.32), (7.2.33) are not identical unless the process is stationary (in which case both $\{g_t(u)\}$ and $\{k_t(v)\}$ are independent of t), but that one set of equations still implies the other. However, the triangular form of both (7.2.32) and (7.2.33) means that these systems of equations are ideally suited for numerical solution.

Continuous-parameter processes

The results for continuous-parameter processes are very similar to those obtained in the discrete case. Thus we are given $\{X(s), -\infty<s\le t\}$ and wish to predict $X(t+m)$ $(m>0)$ by linear combination of past values of the form

$$\tilde{X}(t+m)=\int_{-\infty}^{t} X(s)b_t(s)\,ds, \tag{7.2.34}$$

or, equivalently,

$$\tilde{X}(t+m)=\int_{-\infty}^{t} \xi(u)a_t(u)\,du. \tag{7.2.35}$$

Assuming that $X(t)$ has a linear representation of the form (7.2.22), it follows immediately that

$$\begin{aligned}\tilde{X}(t+m)&=\int_{-\infty}^{t} g_{t+m}(t+m-u)\xi(u)\,du\\ &=\int_{m}^{\infty} g_{t+m}(u)\xi(t+m-u)\,du.\end{aligned} \tag{7.2.36}$$

In order to express $\tilde{X}(t+m)$ in terms of $\{X(s), s\le t\}$ we write

$$\xi(t)=\int_{0}^{\infty} k_t(v)X(t-v)\,dv, \tag{7.2.37}$$

and substitute (7.2.37) in (7.2.22). We then obtain

$$\int_0^v g_t(u)k_{t-u}(v-u)\,du = \delta(v) \tag{7.2.38}$$

for all $v>0$ and all t, where $\delta(v)$ denotes the δ-function. Similarly, substituting (7.2.22) in (7.2.37), we find

$$\int_0^u k_t(v)g_{t-v}(u-v)\,dv = \delta(u) \tag{7.2.39}$$

for all $u \geq 0$ and all t. Given the functions $\{g_t(u)\}$ (for all t), the functions $\{k_t(v)\}$ may, in principle, be determined from either of the above integral equations.

In certain cases the time-domain approach may be used to obtain an expression of the form (7.2.26) or (7.2.34) for $\tilde{X}_{t+m}$, i.e. an expression in terms of past values of X_t, without solving either of the systems (7.2.32) or (7.2.38) directly. Such cases include uniformly modulated processes, and, naturally enough, finite-order autoregressive models (with time-dependent coefficients) can be dealt with also by this method, but they are more suited to a frequency-domain approach (see Section 7.2.2). We now discuss some specific examples.

Example 1: Uniformly modulated process. Consider a continuous-parameter process of the form

$$X(t) = C(t)Y(t),$$

where $Y(t)$ is a stationary process with spectral density function $h_Y(\omega)$, say, and $C(t)$ is some given function of t. Then, with respect to the family $\mathcal{F} \equiv \{e^{i\omega t}C(t)\}$, the evolutionary spectrum of $\{X(t)\}$ is given by

$$h_t(\omega) = |C(t)|^2 h_Y(\omega).$$

Assume that $h_t(\omega)$ satisfies condition C_4. Then it follows that

$$\int_{-\infty}^{\infty} \{\log h_Y(\omega)/(1+\omega^2)\}\,d\omega > -\infty,$$

so that $Y(t)$ has a linear representation of the form

$$Y(t) = \int_0^\infty g(u)\xi(t-u)\,du.$$

Hence we may write $X(t)$ in the form (7.2.22), i.e.

$$X(t) = \int_0^\infty g_t(u)\xi(t-u)\,du,$$

where $g_t(u) = C(t)g(u)$.

According to (7.2.36), the linear least-squares predictor of $X(t+m)$ is given by

$$\begin{aligned}\tilde{X}(t+m) &= \int_m^\infty C(t+m)g(u)\xi(t+m-u)\,du\\ &= C(t+m)\tilde{Y}(t+m),\end{aligned}$$

where $\tilde{Y}(t+m)$ is the least-square predictor $Y(t+m)$, given $\{Y(s), s \le t\}$, and may be determined in the usual way. [This result has a more general form, namely if

$$X(t) = \sum_{u=0}^{\infty} a_t(u)Y(t-u), \quad \text{then } \tilde{X}(t+m) = \sum_{u=0}^{\infty} a_{t+m}(u)Y^*(t+m-u),$$

where

$$Y^*(s) = \begin{cases} Y(s), & s \le t, \\ \tilde{Y}(s), & s > t).] \end{cases}$$

For example, using the Kolmogorov predictor (cf. (7.1.27)),

$$\tilde{Y}(t+m) = \int_0^\infty Y(t-u)b(u)\,du,$$

where

$$b(u) = (2\pi)^{-1}\int_{-\infty}^{\infty} e^{i\omega u}B(\omega)\,d\omega$$

and

$$B(\omega) = \left\{e^{i\omega m}\int_m^\infty g(u)e^{-i\omega u}\,du\right\}\bigg/\int_0^\infty g(u)e^{-i\omega u}\,du.$$

Thus

$$\tilde{X}(t+m) = \int_0^\infty X(t-u)b^*(u)\,du,$$

where

$$b^*(u) = \{C(t+m)/C(t-u)\}\left\{(2\pi)^{-1}\int_{-\infty}^{\infty} e^{i\omega u}B(\omega)\,d\omega\right\}.$$

Example 2: First-order autoregressive process; discrete-parameter. Suppose $\{X_t\}$ is given by

$$X_t - \alpha(t)X_{t-1} = \xi_t, \qquad t = 0, \pm 1, \pm 2, \ldots, \tag{7.2.40}$$

where $\{\alpha(t)\}$ is a given sequence with $|\alpha(t)|<1$, for all t, and $\{\xi_t\}$ is an uncorrelated stationary process with $E\{\xi_t\}=0$, $E\{\xi_t^2\}=1$. From (7.2.40) we find (subject to $X(-\infty)=0$)

$$X_t = \sum_{u=0}^{\infty} g_t(u)\xi_{t-u}$$

where

$$g_t(u)=\begin{cases}\alpha(t)\alpha(t-1)\dots\alpha(t-u+1), & u>0,\\ 1, & u=0.\end{cases} \tag{7.2.41a}$$

Hence, according to (7.2.28), the linear least-squares predictor of X_{t+m} is given by

$$\begin{aligned}\tilde{X}_{t+m} &= \sum_{u=m}^{\infty} \alpha(t+m)\alpha(t+m-1)\dots\alpha(t+m-u+1)\xi_{t+m-u}\\ &=\{\alpha(t+m)\dots\alpha(t+1)\}\sum_{u=0}^{\infty}\alpha(t)\dots\alpha(t-u+1)\xi_{t-u}\\ &=\{\alpha(t+m)\dots\alpha(t+1)\}X_t.\end{aligned} \tag{7.2.41b}$$

(Note that (7.2.41b) is the *exact* expression for X_{t+m} when $\xi_t\equiv 0$, all t.)

Example 3: First-order moving-average process; discrete parameter. Let $\{X_t\}$ be given by

$$X_t=\xi_t-\alpha(t)\xi_{t-1}, \qquad t=0, \pm1, \pm2, \dots, \tag{7.2.42}$$

with $\{\xi_t\}$ as above.

Then clearly,

$$\tilde{X}_{t+m}=\begin{cases}-\alpha(t+1)\xi_t, & m=1,\\ 0, & m>1,\end{cases}$$

where

$$\xi_t=\sum_{v=0}^{\infty} k_t(v)X_{t-v},$$

with $k_t(v)=\alpha(t)\alpha(t-1)\dots\alpha(t-v+1)$.

7.2.2 The Frequency-domain Approach

We now consider an alternative approach based on a generalization of the Wiener–Hopf technique for dealing with the prediction of stationary processes. We treat first the continuous-parameter case, and write the predictor of $X(t+m)$ in the form

$$\tilde{X}(t+m)=\int_0^{\infty} b_t(u)X(t-u)\,du. \tag{7.2.43}$$

Note that (7.2.43) is merely a formal way of writing

$$\tilde{X}(t+m)=\int_0^{\infty} X(t-u)\,d\beta_t(u).$$

(The function $b_t(u)$ need not exist in the strict sense, as would be the case if, for example, $\tilde{X}(t+m)$ involved derivatives of $X(t-u)$.) Note also that the function $b_t(u)$ may depend on both t and m. Now using the evolutionary spectral representation of $X(t)$, (7.2.43) may be written

$$\tilde{X}(t+m)=\int_{-\infty}^{\infty} e^{i\omega t}B_t(\omega)\,dz(\omega), \tag{7.2.44}$$

where

$$B_t(\omega)=\int_0^{\infty} b_t(u)\alpha_{t-u}(\omega)e^{-iu\omega}\,du. \tag{7.2.45}$$

On the other hand, from (7.2.18)

$$X(t+m)=\int_{-\infty}^{\infty} e^{i\omega(t+m)}\alpha_{t+m}(\omega)\,dz(\omega).$$

Hence, using the orthogonality of $z(\omega)$,

$$\begin{aligned}\mathscr{M}(m)&=E[\{\tilde{X}(t+m)-X(t+m)\}^2]\\&=\int_{-\infty}^{\infty} |e^{i\omega m}\alpha_{t+m}(\omega)-B_t(\omega)|^2 d\omega.\end{aligned} \tag{7.2.46}$$

We must now choose $b_t(u)$ (or, equivalently, $B_t(\omega)$) so as to minimize (7.2.46). As a preliminary step we first show that, for each t, the function $B_t(\omega)$ is a "backward transform," that is it may be written as a one-sided Fourier transform of the form

$$B_t(\omega)=(2\pi)^{-1}\int_{-\infty}^{\infty} e^{-iv\omega}K_t(v)\,dv, \tag{7.2.47}$$

where $K_t(v)=0$ $(v<0)$. For

$$\begin{aligned}K_t(v)&=\int_{-\infty}^{\infty} e^{i\omega v}B_t(\omega)\,d\omega\\&=\int_0^{\infty} b_t(u)\left\{\int_0^{\infty} e^{i\omega(v-u)}\alpha_{t-u}(\omega)\,d\omega\right\}du.\end{aligned}$$

Since

$$\alpha_t(\omega) = \int_0^\infty e^{-i\omega u} g_t(u)\, du,$$

$$\int_{-\infty}^{\infty} e^{i\omega(v-u)} \alpha_{t-u}(\omega)\, d\omega = \begin{cases} g_{t-u}(v-u), & u \le v, \\ 0, & u > v. \end{cases}$$

Hence

$$K_t(v) = \int_0^v b_t(u) g_{t-u}(v-u)\, du, \qquad (7.2.48)$$

and, in particular, $K_t(v) = 0$ $(v < 0)$.

Now write for each t,

$$e^{i\omega m} \alpha_{t+m}(\omega) = C_t^{(1)}(\omega) + C_t^{(2)}(\omega),$$

where, for each t, $C_t^{(1)}(\omega)$ is a "backward transform", and $C_t^{(2)}(\omega)$ is a "forward transform", i.e. we write

$$C_t^{(1)}(\omega) = (2\pi)^{-1} \int_0^\infty e^{-i\omega u} \left[\int_{-\infty}^{\infty} e^{i\theta u} \{ e^{i\theta m} \alpha_{t+m}(\theta) \}\, d\theta \right] du$$

$$C_t^{(2)}(\omega) = (2\pi)^{-1} \int_{-\infty}^{0} e^{-i\omega u} \left[\int_{-\infty}^{\infty} e^{i\theta u} \{ e^{i\theta m} \alpha_{t+m}(\theta) \}\, d\theta \right] du.$$

Then, for each t, $\{C_t^{(1)}(\omega) - B_t(\omega)\}$ and $C_t^{(2)}(\omega)$ are orthogonal with respect to ω-integration, and from (7.2.46) we have

$$\mathcal{M}(m) = \int_{-\infty}^{\infty} |C_t^{(1)}(\omega) - B_t(\omega)|^2 d\omega + \int_{-\infty}^{\infty} |C_t^{(2)}(\omega)|^2\, d\omega. \qquad (7.2.49)$$

It follows immediately that the minimum of $\mathcal{M}(m)$ is attained when $B_t(\omega)$ is given by

$$B_t(\omega) = C_t^{(1)}(\omega), \qquad \text{all } t, \qquad (7.2.50)$$

and the prediction variance is then given by

$$\mathcal{M}(m) = \int_{-\infty}^{\infty} |C_t^{(2)}(\omega)|^2\, d\omega. \qquad (7.2.51)$$

It now remains to solve (7.2.50) to find the function $\{b_t(u)\}$. Taking Fourier transforms of both sides we find, for $v \ge 0$,

$$K_t(v) = \int_{-\infty}^{\infty} e^{i\theta(v+m)} \alpha_{t+m}(\theta)\, d\theta \qquad \text{for all } t,$$

$$= g_{t+m}(v+m).$$

Hence, using (7.2.48), we find

$$\int_0^v b_t(u) g_{t-u}(v-u)\, du = g_{t+m}(v+m), \qquad (7.2.52)$$

for all t and all $v \geq 0$. Given the form of the evolutionary spectral density function, $h_t(\omega)$, for all t, we may (in principle) determine the function $g_t(u)$, for all t, and hence find the optimum function $\{b_t(u)\}$ by solving the integral equation (7.2.52).

Equation (7.2.52), being an integral equation of the first kind, may be solved explicitly for $\{b_t(u)\}$. (Note that in the stationary case $h_t(\omega)$, and consequently $g_t(u)$, are both independent of t, so that (7.2.52) may then be solved immediately, by noting that the LHS is the convolution of $b_t(u)$ and $g_t(u)$.) In the general case we may solve (7.2.52) by first expressing it as an integral equation of the second kind. Thus, if we assume

(i) $g_{t-v}(0) \neq 0$ for all v in $(0, \infty)$ and
(ii) $\partial g_{t-u}(v-u)/\partial v = g'_{t-u,v}(v-u)$ (say) exists and is continuous in u and v $(u \leq v)$, then by differentiating both sides of (7.2.52) w.r. to v we find

$$g_{t-v}(0) b_t(v) + \int_0^v g'_{t-u,v}(u-v) b_t(u)\, du = g'_{t+m,v}(v+m) \qquad (v \geq 0)$$

or

$$b_t(v) + \int_0^v \frac{g'_{t-u,v}(v-u)}{g_{t-v}(0)} b_t(u)\, du = \frac{g'_{t+m,v}(v+m)}{g_{t-v}(0)}. \qquad (7.2.53)$$

The solution of (7.2.53) is (see Tricomi, 1957, p. 10)

$$b_t(v) = \{g'_{t+m,v}(v+m)/g_{t-v}(0)\}$$
$$+ \int_0^v H_t^*(v, u, -1)\{g'_{t+m,v}(u+m)/g_{t-u}(0)\}\, du, \qquad (7.2.54)$$

where, for each t, the "resolvent kernel", $H_t^*(v, u, -1)$, is given by

$$H_t^*(v, u, -1) = -\sum_{p=0}^{\infty} (-1)^p K_{t,p+1}^*(u, v)$$

and the "iterated kernels", $K_{t,p}^*(u,v)$, are given by

$$K_{t,p+1}^*(u, v) = \int_u^v K_{t,1}^*(u, z) K_{t,p}^*(z, v)\, dz, \qquad p = 1, 2, \ldots,$$

with

$$K_{t,1}^*(u, v) = \{g'_{t-u,v}(v-u)/g_{t-v}(0)\}.$$

Equation (7.2.54) provides an analytic expression for $\{b_t(u)\}$, but in numerical work it may well be more convenient to solve the integral equation (7.2.52) directly.

For discrete-parameter processes the predictor $\tilde{X}_{t+m}$ will be of the form

$$\tilde{X}_{t+m} = \sum_{u=0}^{\infty} b_t(u) X_{t-u} \tag{7.2.55}$$

so that, using the spectral representation,

$$\mathcal{M}(m) = \int_{-\pi}^{\pi} |e^{i\omega m} \alpha_{t+m}(\omega) - B_t(\omega)|^2 \, d\omega, \tag{7.2.56}$$

where now

$$B_t(\omega) = \sum_{u=0}^{\infty} b_t(u) \alpha_{t-u}(\omega) \, e^{-iu\omega}.$$

Following a similar argument to that used above, we may show that $\mathcal{M}(m)$ is minimized when the sequence $\{b_t(u)\}$ is given by

$$\sum_{u=0}^{v} b_t(u) g_{t-u}(v-u) = g_{t+m}(v+m), \qquad v = 0, 1, 2, \ldots; \, t = 0, \pm 1, \pm 2, \ldots. \tag{7.2.57}$$

Here, $\{g_t(u)\}$ $(u = 0, 1, 2, \ldots)$ is the sequence defined by (7.2.9).

The system of equations (7.2.57) is again "triangular" so that it is easily solved for $\{b_t(u)\}$ by repeated back-substitution. In fact, this approach has been tried on a numerical example and has proved quite successful.

It is interesting to note that had we attempted to obtain $\{b_t(u)\}$ directly from the covariance function, $R(s, t) \equiv E\{X(s)X^*(t)\}$, we would have obtained the well-known set of equations (cf. (7.1.31))

$$\sum_{u=0}^{\infty} b_t(u) R(t-u, t-v) = R(t-v, t+m), \qquad v \geq 0. \tag{7.2.58}$$

The above system does not lend itself to either an analytic or a numerical solution, and the spectral approach may be regarded as a method of reducing (7.2.58) to "triangular" form.

We now apply the methods of this section to some examples discussed in Section 7.1.1.

Example 4. Consider the first-order autoregressive process (discrete-parameter). Using the form of $\{g_t(u)\}$ given by (7.2.41a) in (7.2.57), we obtain

$$b_t(0) + \sum_{u=1}^{v} b_t(u) \alpha(t-u) \alpha(t-u-1) \ldots \alpha(t-v+1)$$
$$= \alpha(t+m) \alpha(t+m-1) \ldots \alpha(t-v+1), \qquad v = 0, 1, 2, \ldots.$$

Solving this system step by step, we obtain

$$\begin{cases} b_t(0) = \alpha(t+m)\alpha(t+m-1)\ldots\alpha(t+1) \\ b_t(u) = 0, \qquad u > 0. \end{cases}$$

Hence

$$\tilde{X}(t+m) = \alpha(t+m)\alpha(t+m-1)\ldots\alpha(t+1)X(t),$$

in agreement with Example 2.

Example 5. Consider now the moving-average process, discrete parameter. For this process,

$$g_t(u) = \begin{cases} 1, & u = 0, \\ -\alpha(t), & u = 1, \\ 0, & u > 1, \end{cases}$$

for all t. Substituting this expression in (7.2.57), we find

$$b_t(0) = \begin{cases} -\alpha(t+1), & m = 1, \\ 0, & m > 1, \end{cases}$$

$$b_t(u)g_{t-u}(1) + b_t(u+1)g_{t-u}(0) = 0, \qquad u = 1, 2, \ldots, \text{ all } m.$$

The solution of this set of equations is, for all u,

$$b_t(u) = \begin{cases} -\alpha(t+1)\alpha(t)\ldots\alpha(t-u+1), & m = 1, \\ 0, & m > 1, \end{cases}$$

so that

$$\tilde{X}(t+m) = \begin{cases} -\sum_{u=0}^{\infty} \alpha(t+1)\alpha(t)\ldots\alpha(t-u+1)X(t), & m = 1, \\ 0, & m > 1, \end{cases}$$

in agreement with Example 3.

Discussion

As mentioned previously, we are here concerned primarily with the problem of determining the optimum predictor on the assumption that the second-order properties of the process are fully known. However, we have shown that the predictor, $\tilde{X}(t+m)$, is determined uniquely by the form of the evolutionary spectra, $\{h_t(\omega)\}$, so that in cases where we have no precise knowledge of the underlying model we may attack the problem by first estimating the functions $\{h_t(\omega)\}$. Now it will be seen that $\tilde{X}(t+m)$ depends on the form of $g_{t+m}(u)$, i.e. on the form of the evolutionary spectrum at the *future* time point $(t+m)$. In a purely empirical analysis we would not, of course, be able to estimate $h_{t+m}(\omega)$ directly from observations extending only up to time t, and we would have to assume that the spectra were changing with sufficient "smoothness" over time to enable the form of $h_{t+m}(\omega)$ to be inferred from the spectra which have been estimated up to time t. If the prediction step m is much smaller than the characteristic width, B_x, of the process, it would be a reasonable approximation to replace $h_{t+m}(\omega)$ by $h_t(\omega)$—the most recently available spectrum. In fact, it seems clear intuitively that the very nature of a non-stationary process precludes long-range prediction, unless we are prepared to make assumptions about the character of the non-stationarity. Consequently, if $m \geq B_x$, we would have to extrapolate the available spectra to the time point $(t+m)$. This extrapolation could be performed on the spectral ordinates themselves (by fitting suitable functions of t to a range of values of ω), but in general we would presumably start from a model of the process which determines the form of the evolutionary spectra in terms of a set of time-varying parameters, $\{\alpha_i(t)\}$, say, and then extrapolate the values of these parameters. (Thus, for the model of Example 2, the $\{h_t(\omega)\}$ may be expressed in terms of the time-varying parameter $\alpha(t)$.) If our model specifies further the functional time-dependence of the $\{\alpha_i(t)\}$ in terms of another set of constant parameters, $\{\beta_j\}$, say, then the extrapolation is a straightforward problem and involves merely the estimation of the $\{\beta_j\}$. (For example, in some cases we may postulate that the $\{\alpha_i(t)\}$ are periodic functions of t with known periodicities.) If, on the other hand, the functional form of the $\{\alpha_i(t)\}$ is not specified by the model, the extrapolation would have to be performed by regression techniques.

7.3 FILTERING OF NON-STATIONARY PROCESSES

The basic nature of the filtering problem may be described as follows: we are able to observe a (possibly non-stationary) process $\{U(t)\}$ which may be written in the form

$$U(t) = S(t) + N(t),$$

where $\{S(t)\}$, $\{N(t)\}$ have known second-order properties. Given a realization of U extending up to time t, we wish to approximate to $S(t+m)$ by using a linear combination (or "filter") of $\{U(s); s \leq t\}$. Here, m may be positive, negative or zero, and the "coefficients" of the linear combination are chosen so as to minimize the mean-square error of the approximation. When $N(t) \equiv 0$ and $m > 0$, the problem reduces to pure prediction, but in the language of communication engineering we would describe $\{S(t)\}$ as the "signal" and $\{N(t)\}$ as the "noise", so that the general problem may be interpreted as the "extraction of a signal from a record of signal plus noise". The case of stationary $\{S(t)\}$ would correspond, for instance, to a "stable" radio signal, and a non-stationary $\{S(t)\}$ to a "fading" or "wandering" signal. (For a discussion of the stationary filtering problem see, for example, *Spec. Anal.*, p. 773; Whittle, 1963.) In our model we allow both $\{S(t)\}$ and $\{N(t)\}$ to be non-stationary, so that we allow for the possibility that the "noise" also possesses non-stationary characteristics. (Hannan (1967), using a different approach, has discussed a particular case of this problem; namely when the non-stationary character of $\{S(t)\}$ arises from an autoregressive model of the "explosive" type.) The approach described below is due to Abdrabbo and Priestley (1969).

Optimum filtering

Consider first the case of continuous-parameter processes and write the approximation of $S(t+m)$ in the form

$$\hat{S}(t+m) = \int_0^\infty b_t(v)U(t-v)\,dv, \tag{7.3.1}$$

remembering that, in the above, m may be positive, negative, or zero. (The case $m \leq 0$ corresponds to "smoothing out" the noise component from a portion of the record already observed, and the case $m > 0$ corresponds to "predicting" a future value of the signal.) The function $\{b_t(u)\}$ is to be chosen so that

$$\mathcal{M}(m) = E[\{\hat{S}(t+m) - S(t+m)\}^2] \tag{7.3.2}$$

is minimized.

Assume now that $S(t)$ and $N(t)$ are uncorrelated processes, and that they admit *evolutionary spectral* representations of the form (6.3.20) with respect to a common family $\mathcal{F} = \{e^{it\omega}A_t(\omega)\}$ and measures $\mu_S(d\omega)$, $\mu_N(d\omega)$, respectively, each of which is absolutely continuous with respect to $d\omega$. Then we may write

$$U(t) = \int_{-\infty}^\infty e^{it\omega}A_t(\omega)\,dZ_U(\omega), \tag{7.3.3}$$

where

$$E\{|dZ_U(\omega)|^2\} = \mu_U(d\omega) = \mu_S(d\omega) + \mu_N(d\omega). \tag{7.3.4}$$

Using (7.3.3), equation (7.3.1) may be written

$$\hat{S}(t+m) = \int_{-\infty}^\infty e^{i\omega t}\beta_t(\omega)\,dZ_U(\omega), \tag{7.3.5}$$

where

$$\beta_t(\omega) = \int_0^\infty b_t(v) A_{t-v}(\omega)\, e^{-i\omega v}\, dv. \tag{7.3.6}$$

However, we have also

$$S(t+m) = \int_{-\infty}^{\infty} e^{i\omega(t+m)} A_{t+m}(\omega)\, dZ_S(\omega), \tag{7.3.7}$$

so that

$$\mathcal{M}(m) = E\left| \int_{-\infty}^{\infty} e^{i\omega(t+m)}\, A_{t+m}(\omega)\, dZ_S(\omega) - \int_{-\infty}^{\infty} e^{i\omega t} \beta_t(\omega)\, dZ_U(\omega) \right|^2. \tag{7.3.8}$$

We now assume further that both $S(t)$ and $U(t)$ possess linear representations of the type discussed in Section 7.2, so that, with an obvious notation, we obtain the factorizations,

$$h_{SS,t}(\omega) = |A_t(\omega)|^2 (d\mu_S/d\omega) = |\alpha_t^{(S)}(\omega)|^2; \quad \text{say,}$$

$$h_{UU,t}(\omega) = |A_t(\omega)|^2 (d\mu_U/d\omega) = |\alpha_t^{(U)}(\omega)|^2, \quad \text{say,}$$

where both $\alpha_t^{(S)}(\omega)$ and $\alpha_t^{(U)}(\omega)$ have one-sided Fourier transforms. Then we may rewrite equation (7.3.8) as

$$\mathcal{M}(m) = \int_{-\infty}^{\infty} |e^{im\omega} \alpha_{t+m}^{(0)}(\omega) - B_t(\omega)|^2\, d\omega$$
$$+ \int_{-\infty}^{\infty} h_{SS,t+m}(\omega) \left\{ 1 - \left(\frac{d\mu_S}{d\omega} \Big/ \frac{d\mu_U}{d\omega} \right) \right\} d\omega. \tag{7.3.9}$$

Here,

$$\alpha_t^{(0)}(\omega) = \{ h_{SS,t}(\omega) / \alpha_t^{*(U)}(\omega) \} \tag{7.3.10}$$

(* denoting the complex conjugate) and

$$B_t(\omega) = \int_0^\infty b_t(v) \alpha_{t-v}^{(U)}(\omega)\, e^{-iv\omega}\, dv. \tag{7.3.11}$$

We have now to choose $b_t(v)$ (or equivalently $B_t(\omega)$) so as to minimize (7.3.9).

As in the case of the prediction problem we can show that, for each t, $B_t(\omega)$ is a "backward transform". More precisely, if we write, for each t,

$$B_t(\omega) = \frac{1}{2\pi}\int_{-\infty}^{\infty} e^{-iv\omega} K_t(v)\,dv,$$

then

$$K_t(v) = \begin{cases} \displaystyle\int_0^v b_t(u) g_{t-u}^{(U)}(v-u)\,du, & v \ge 0, \\ 0, & v < 0. \end{cases} \tag{7.3.12}$$

In the above $g_t^{(U)}(v)$ is the Fourier transform of $\alpha_t^{(U)}(\omega)$, i.e.

$$\alpha_t^{(U)}(\omega) = \int_0^{\infty} e^{-i\omega v} g_t^{(U)}(v)\,dv. \tag{7.3.13}$$

The form of $B_t(\omega)$ which minimizes (7.3.9) is again obtained by the technique of decomposing $\{e^{im\omega}\ \alpha_{t+m}^{(0)}(\omega)\}$ into the sum of "backward" and "forward" transforms. Thus we write (for each t),

$$e^{im\omega}\alpha_{t+m}^{(0)}(\omega) = C_{t+m}^{(1)}(\omega) + C_{t+m}^{(2)}(\omega), \tag{7.3.14}$$

where $C_{t+m}^{(1)}(\omega)$ (the "backward" transform) is given by

$$C_{t+m}^{(1)}(\omega) = \frac{1}{2\pi}\int_0^{\infty} e^{-i\omega u}\left[\int_{-\infty}^{\infty} e^{i\theta u}\{e^{i\theta m}\alpha_{t+m}^{(0)}(\theta)\}\,d\theta\right] du \tag{7.3.15}$$

and $C_{t+m}^{(2)}(\omega)$ (the "forward" transform) is given by

$$C_{t+m}^{(2)}(\omega) = \frac{1}{2\pi}\int_{-\infty}^{0} e^{-i\omega u}\left[\int_{-\infty}^{\infty} e^{i\theta u}\{e^{i\theta m}\alpha_{t+m}^{(0)}(\theta)\}\,d\theta\right] du. \tag{7.3.16}$$

Equation (7.3.9) now becomes

$$\mathscr{M}(m) = \int_{-\infty}^{\infty} |C_{t+m}^{(2)}(\omega)|^2\,d\omega + \int_{-\infty}^{\infty} |C_{t+m}^{(1)}(\omega) - B_t(\omega)|^2 d\omega$$
$$+ \int_{-\infty}^{\infty} h_{SS,t+m}(\omega)\left\{1 - \left(\frac{d\mu_S}{d\omega}\right)\Big/\left(\frac{d\mu_U}{d\omega}\right)\right\} d\omega \tag{7.3.17}$$

and $\mathscr{M}(m)$ is clearly a minimum when

$$B_t(\omega) = C_{t+m}^{(1)}(\omega). \tag{7.3.18}$$

Taking Fourier transforms of both sides of (7.3.18) we obtain, for $v \ge 0$ (using (7.3.12)),

$$\int_0^v b_t(u) g_{t-u}^{(U)}(v-u)\,du = l_{t+m}(v), \tag{7.3.19}$$

where $l_{t+m}(v)$, the Fourier transform $C^{(1)}_{t+m}(\omega)$ is given by

$$l_{t+m}(v) = \int_{-\infty}^{\infty} C^{(1)}_{t+m}(\omega) e^{i\omega v}\, d\omega$$
$$= \int_{-\infty}^{\infty} e^{i\theta(v+m)} \alpha^{(0)}_{t+m}(\theta)\, d\theta. \tag{7.3.20}$$

With the above choice for $B_t(\omega)$, the minimum mean-square filtering error is

$$\mathcal{M}_{\min}(m) = \int_{-\infty}^{\infty} |C^{(2)}_{t+m}(\omega)|^2\, d\omega + \int_{-\infty}^{\infty} h_{SS,t+m}(\omega) \left\{1 - \left(\frac{d\mu_S}{d\omega}\right) \Big/ \left(\frac{d\mu_U}{d\omega}\right)\right\} d\omega \tag{7.3.21}$$

Note that since $\mu_U = \mu_S + \mu_N$, the second term of (7.3.21) may be expressed in the alternative form

$$\int_{-\infty}^{\infty} \{(h_{SS,t+m}(\omega) h_{NN,t+m}(\omega))/(h_{SS,t+m}(\omega) + h_{NN,t+m}(\omega))\}\, d\omega. \tag{7.3.22}$$

The result is completely analogous to that stated in Yaglom (1962, p. 133), with "stationary" spectra replaced by evolutionary spectra. (Note also that in (7.3.22) we have, for convenience, expressed all evolutionary density functions in terms of their values at the time instant $(t+m)$. However, apart from the factor in the numerator, we could replace $(t+m)$ by *any* other value of t—say, $t=0$.) Yaglom points out that the expression (7.3.22)—or rather its analogue for the stationary case—may be interpreted as the minimum mean-square error for the case when one observed $U(t)$ for all t (that is, an infinite realization), in which case $B_t(\omega)$ need no longer be restricted to the class of "backward" transforms. He remarks that it is impossible, therefore, to recover $S(t)$ with perfect precision unless the product of the "signal" and "noise" spectra vanishes at all frequencies; that is, unless the "signal" and "noise" have non-overlapping spectra. (As Yaglom further observes, this result is fairly obvious intuitively if one thinks of the filtering operation in terms of "band-pass" filters.) It is interesting to note that a corresponding result holds for non-stationary processes; that is, that the expression (7.3.22) vanishes only when the "signal" and "noise" have *non-overlapping evolutionary spectra.* However, we do not require the two sets of evolutionary spectra to be non-overlapping at all time instants; by virtue of the remarks noted above, (7.3.22) will vanish if for each t, $h_{SS,t}(\omega)$ does not overlap with (say) $h_{NN,0}(\omega)$, or equivalently, if for each t, $h_{NN,t}(\omega)$ does not overlap with (say) $h_{SS,0}(\omega)$. That a result of this type holds is obviously a consequence of our initial assumption that $S(t)$ and $N(t)$ admit representation with respect to the same family of functions. It might be conjectured that in a more general situation the expression (7.3.22) would still represent the mean-square error for an infinite realization, and that consequently $S(t)$ could still be determined with perfect precision provided $h_{SS,t}(\omega)$ and $h_{NN,t}(\omega)$ are non-overlapping for each value of t.

Discrete-parameter processes

The problem for discrete-parameter processes may be treated similarly. Thus, we consider a filtering formula of the form

$$\hat{S}_{t+m} = \sum_{v=0}^{\infty} b_t(v) U_{t-v} \tag{7.3.23}$$

and choose $\{b_t(v)\}$ so as to minimize

$$\begin{aligned}\mathcal{M}(m) = &\int_{-\pi}^{\pi} |e^{im\omega}\alpha_{t+m}^{(0)}(\omega) - B_t(\omega)|^2 d\omega \\ &+ \int_{-\pi}^{\pi} h_{SS,t+m}(\omega)\left\{1 - \left(\frac{d\mu_S}{d\omega}\right) \Big/ \left(\frac{d\mu_U}{d\omega}\right)\right\} d\omega.\end{aligned} \tag{7.3.24}$$

Here,

$$B_t(\omega) = \sum_{v=0}^{\infty} b_t(v)\alpha_{t-v}^{(U)}(\omega)\, e^{-i\omega v}. \tag{7.3.25}$$

Following a similar analysis to that used in the continuous case we derive a set of equations analogous to (7.3.19), namely

$$\sum_{u=0}^{v} b_t(u) g_{t-u}^{(U)}(v-u) = l_{t+m}(v), \qquad v \geq 0. \tag{7.3.26}$$

As in the case of pure prediction, the above system is "triangular" and may be solved iteratively.

Filtering a uniformly modulated process

Consider the case where S_t and N_t are both discrete parameters, uniformly modulated processes of the form

$$S_t = G(t) X_t^{(0)}$$

$$N_t = G(t) Y_t^{(0)}$$

where $G(t)$ is some function of t only, $\{Y_t^{(0)}\}$ is an uncorrelated (stationary) process with $E[Y_t^{(0)}] = 0$, $E[\{Y_t^{(0)}\}^2] = \sigma_y^2$, and $X_t^{(0)}$ is a (stationary) AR(1) process of the form

$$X_t^{(0)} - aX_{t-1}^{(0)} = Z_t^{(0)},$$

in which $\{Z_t^{(0)}\}$ is an uncorrelated (stationary) process with $E[Z_t^{(0)}]=0$, $E[\{Z_t^{(0)}\}^2]=\sigma_z^2$.

With respect to the family $\mathscr{F}\equiv\{e^{i\omega t}G(t)\}$, the evolutionary spectrum of $U_t=S_t+N_t$ is given by (cf. Yaglom, 1962, p. 60),

$$h_{UU,t}(\omega)=B^2|G(t)|^2\left|\frac{e^{i\omega}-b}{e^{i\omega}-a}\right|^2, \qquad (B^2=C/2\pi),$$

where C and b are real numbers ($C>0$, $|b|<1$) determined by the equations

$$\frac{Cb}{a}=\sigma_z^2, \qquad \frac{C(a-b)(1-ab)}{a(1-a^2)}=\sigma_y^2.$$

In this case we obtain

$$\alpha_t^{(0)}(\omega)=\frac{KG(t)}{(1-ab)}\left\{\frac{1}{(1-ae^{-i\omega})}+\frac{be^{i\omega}}{(1-be^{i\omega})}\right\},$$

where

$$K=[\{\sigma_y^2(1-a^2)\}/(2\pi B)],$$

and

$$B_t(\omega)=\left\{\sum_{v-0}^{\infty}b_t(v)G(t-v)e^{-i\omega v}\right\}\left\{\frac{B(e^{i\omega}-b)}{(e^{i\omega}-u)}\right\}$$

$$=\beta_t(\omega)\left[\frac{B(e^{i\omega}-b)}{(e^{i\omega}-a)}\right], \quad \text{say.}$$

For this type of process the Fourier coefficients of $B_t(\omega)$ are readily seen to be proportional to $\{b_t(v)G(t-v)\}$, so that it is simpler to determine $\{b_t(v)\}$ from the discrete analogue of (7.3.18) rather than (7.3.19). Consider first the case $m\geq 0$. We find

$$C_{t+m}^{(1)}(\omega)=\frac{Ka^mG(t+m)}{(1\quad ab)}\sum_{v=0}^{\infty}e^{-i\omega v}a^v.$$

From the discrete analogue of equation (7.3.18) we now obtain

$$\beta_t(\omega)=C_{t+m}^{(1)}(\omega)\left\{\frac{e^{i\omega}-a}{B(e^{i\omega}-b)}\right\}=\frac{Ka^mG(t+m)}{B(1-ab)}\sum_{v=0}^{\infty}e^{-i\omega v}b^v,$$

so that, equating Fourier coefficients,

$$b_t(v)=\frac{G(t+m)}{G(t-v)}\left\{\frac{Ka^m}{B(1-ab)}\right\}b^v, \qquad v\geq 0.$$

Thus, for $m \geq 0$, the filtering equation becomes

$$\hat{S}_{t+m} = G(t+m)(a-b)a^{m-1} \sum_{v=0}^{\infty} \frac{b^v}{G(t-v)} U_{t-v}.$$

For $m < 0$, we find

$$C_{t+m}^{(1)}(\omega) = \frac{KG(t+m)}{(1-ab)} \left[\sum_{v=0}^{-m-1} b^{-m-v} e^{-i\omega v} + \sum_{v=-m}^{\infty} a^{m+v} e^{-i\omega v} \right],$$

so that, after some reduction,

$$\beta_t(\omega) = \frac{(a-b)G(t+m)b^{-m}}{a(1-be^{-i\omega})} \left\{ 1 + (1-ab) \sum_{v=0}^{-m-1} b^{-(v+1)} e^{-i\omega(v+1)} \right\}$$

$$= \sum_{v=0}^{\infty} l_t(v) e^{-i\omega v}, \quad \text{say.}$$

The sequence $\{b_t(v)\}$ is now obtained by equating Fourier coefficients in the above equation, giving

$$b_t(v) = l_t(v)/G(t-v).$$

To illustrate an explicit expression for $\{b_t(v)\}$ we consider the case $m = -1$, which gives

$$b_t(0) = \frac{(a-b)}{a} \frac{G(t-1)}{G(t)} b,$$

$$b_t(v) = \frac{(a-b)}{a} \frac{G(t-1)}{G(t-v)} \{b^{v+1} - (1-ab)b^{v-1}\}, \qquad v \geq 1.$$

Thus we finally obtain

$$\hat{S}_{t-1} = \frac{(a-b)G(t-1)}{a} \left\{ \frac{b}{G(t)} U_t + (b^2 + ab - 1) \sum_{v=1}^{\infty} \frac{b^{v-1}}{G(t-v)} U_{t-v} \right\}.$$

7.4 MINIMUM VARIANCE CONTROL OF SYSTEMS INFECTED BY NON-STATIONARY DISTURBANCES

Once we have developed a theory of prediction for non-stationary processes we may apply these results to linear control systems in which the "noise" disturbance is non-stationary. Consider the system described schematically in Fig. 7.1.

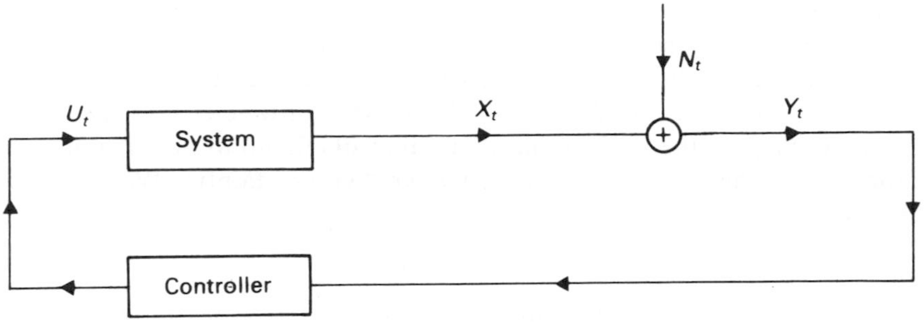

$U_t = f_t(Y_t, Y_{t-1}, \ldots .)$

Fig. 7.1. Feedback controller.

Here, U_t denotes the input (at time t), X_t is the true (unobservable) output, N_t is an additive noise disturbance, and Y_t is the observed output. The problem is to design a controller which, for each t, computes the input U_t as a function of Y_t, $Y_{t-1}, \ldots$ so as to minimize some cost function involving $\{Y_t\}$ and $\{U_t\}$. A typical situation is that in which we wish to control the output Y_t so as to be as close as possible to some predetermined value—say zero. In this case a natural cost function would be $V = E[Y_t^2]$ and accordingly we would have to choose the input so as to minimize $E[Y_t^2]$—leading to the terminology *minimum variance control.*

Suppose now that the system has linear dynamics so that the relationship between U_t and X_t may be expressed in the form

$$X_t = \sum_{u=0}^{\infty} a_u U_{t-u} = A(B)U_t, \tag{7.4.1}$$

where

$$A(z) = \sum_{u=0}^{\infty} a_u z^u.$$

(The function $A(e^{-i\omega})$ is, of course, the system's transfer function.) The observed output Y_t is then given by,

$$Y_t = X_t + N_t = A(B)U_t + N_t. \tag{7.4.2}$$

Suppose further that there is a delay (or "dead time") of d (≥ 0) time units, so that $a_0 = a_1 = \ldots a_{d-1} = 0$, $a_d \neq 0$. In this case the input at time t does not begin to affect the output until time $(t+d)$, and the optimal control input at time t will be that for which $V = E[Y_{t+d}^2]$ is minimized. The solution may be obtained from the principle of *certainty equivalence* (Whittle, 1963, p. 134), or the *separation principle*, as it is known in the control theory literature, and, as noted by Box and Jenkins (1970), it tells us that in order to keep Y_{t+d} as close as possible to zero we should choose the input so that the corresponding output X_{t+d} exactly cancels the best predictor (in the mean-square sense) of the noise disturbance at time $(t+d)$. Thus, we compute $\tilde{N}_{t+d}$ on the basis of information available up to time t, and then proceed as if the value of N_{t+d} were known "with certainty". We therefore choose U_t so that

$$X_{t+d} = -\tilde{N}_{t+d}, \tag{7.4.3}$$

where $\tilde{N}_{t+d}$ is the linear least-squares predictor of N_{t+d} based on information available up to time t. Using (7.4.1), the input U_t may be written formally as

$$U_t = -\{B^d A^{-1}(B)\}\tilde{N}_{t+d}. \tag{7.4.4}$$

Thus, the problem of constructing the optimum control input reduces to that of least-squares prediction.

Box and Jenkins (1970) suggested that the noise disturbance, N_t, may be described by an ARIMA model of the form

$$\alpha(B)(1-B)^D N_t = \beta(B)\varepsilon_t, \tag{7.4.5}$$

where $\{\varepsilon_t\}$ is a white noise process, and α, β are finite-order polynomials of the form

$$\alpha(z) = 1 + a_1 z + \cdots + a_k z^k,$$

$$\beta(z) = 1 + b_1 z + \cdots + b_l z^l.$$

According to this model, N_t may be regarded as the output of an "unstable" filter with transfer function $\beta(z)/\alpha(z)(1-z)^D$ (having a Dth-order pole at $z=1$) when the input is white noise. The process N_t would therefore be non-stationary, but its non-stationary characteristics are determined by a fixed (i.e. time invariant) set of parameters. If we fit an ARIMA model of the form (7.4.5) to an initial set of data, then the predicted future evolution of the process is completely determined by the form of the initial data set. To free ourselves from this restriction we now consider a more general form of non-stationary model in which the parameters are themselves time dependent, thus allowing for an arbitrary type of non-stationary character. This type of model may be written as

$$\alpha_t(B)(1-B)^D N_t = \beta_t(B)\varepsilon_t, \tag{7.4.6}$$

where $\alpha_t(z)$, $\beta_t(z)$ are polynomials of the form

$$\alpha_t(z) = 1 + a_{1,t} z + \cdots + a_{k,t} z^k$$

$$\beta_t(z) = 1 + b_{1,t} z + \cdots + b_{l,t} z^l,$$

and, for each t, the roots of $\alpha_t(z)$, $\beta_t(z)$ lie outside the unit circle. Prediction formulae for this model can be derived by the general method described in Section 7.2, and estimates of the time-dependent parameters $(a_{1,t}, \ldots, a_{k,t}, b_{1,t}, \ldots, b_{l,t})$ can be obtained by fitting a rational function to the estimated evolutionary spectra of $\Delta^D N_t$—see Priestley (1969).

The above approach leads to a form of *parameter-adaptive control* in which the controller sets the input in accordance with the *current* form of the noise process, and will automatically adjust to changes in the noise structure. At each instant of time the input is adjusted so as to be optimal for the local behaviour of the noise process.

As an illustration we consider the following noise model,

$$\Delta^2 N_t = \varepsilon_t + c_{1,t}\varepsilon_{t-1} + c_{2,t}\varepsilon_{t-2} + c_{3,t}\varepsilon_{t-3},$$

so that in terms of the general model (7.4.6),

$$D = 2, \quad \alpha_t(z) \equiv 1, \quad \beta_t(z) = 1 + c_{1,t}z + c_{2,t}z^2 + c_{3,t}z^3.$$

Suppose now that the system has unit delay, i.e. $d = 1$, so that we need to evaluate $\tilde{N}_{t+1}$. It may be shown (Priestley, 1969) that

$$\Delta\tilde{N}_{t+1} = \Delta\{\gamma_{-1}(t)\varepsilon_t\} + \gamma_0(t)\varepsilon_t + S\{\gamma_1(t)\varepsilon_t\}, \tag{7.4.7}$$

where $S = \Delta^{-1}$, denotes the summation operator, and $\gamma_{-1}(t)$, $\gamma_0(t)$, $\gamma_1(t)$, are given by

$$\gamma_{-1}(t) = c_{3,t+3},$$
$$\gamma_0(t) = 1 - 2c_{3,t+3} - c_{2,t+2},$$
$$\gamma_1(t) = c_{1,t+1} + c_{2,t+2} + c_{3,t+3} + 1$$

(cf. the discussion of the corresponding fixed parameter model given in *Spec. Anal.*, p. 784). The control input is then given by

$$\Delta U_t = -\{BA^{-1}(B)\}\Delta\tilde{N}_{t+1}, \tag{7.4.8}$$

and with this input, $Y_t = N_t - \tilde{N}_t = \varepsilon_t$, so that (7.4.8) can be written

$$\Delta U_t = -\{BA^{-1}(B)\}[\Delta\{\gamma_{-1}(t)Y_t\} + \gamma_0(t)Y_t + S\{\gamma_1(t)Y_t\}]. \tag{7.4.9}$$

The controller (7.4.9) may be regarded as an adaptive version of the *three-term controller* discussed by Box and Jenkins (1970). The weights, γ_{-1}, γ_0, γ_1, attached to the "differential", "proportional", and "integral" terms are now allowed to vary over time, their values at each instant being those which are "locally" optimal.

Assuming that $c_{1,t}$, $c_{2,t}$, $c_{3,t}$ are sufficiently smooth functions of t for the family

$$\mathcal{F}_c = \{1 + c_{1,t}e^{-i\omega} + c_{2,t}\,e^{-2i\omega} + c_{3,t}\,e^{-3i\omega}\}$$

to be oscillatory, the evolutionary spectra of $\Delta^2 N_t$ with respect to $\mathscr{F}_c$ are given by,

$$h_t^{(\Delta^2 N)}(\omega) = \frac{\sigma_\varepsilon^2}{2\pi}|1 + c_{1,t}\, e^{-i\omega} + c_{2,t}\, e^{-2i\omega} + c_{3,t}\, e^{-3i\omega}|^2. \qquad (7.4.10)$$

The values of the estimated form of $h_t^{(\Delta^2 N)}(\omega)$ at any three frequencies would then be sufficient to obtain rough estimates of $c_{1,t}$, $c_{2,t}$, $c_{3,t}$, but it would be more efficient to fit the appropriate model to $\log\{h_t^{(\Delta^2 N)}(\omega)\}$ and then estimate $c_{1,t}$, $c_{2,t}$, $c_{3,t}$, by least-squares.

Although in the above discussion we allowed the noise process to have a general non-stationary character, the system dynamics were assumed to be linear and time-invariant. The treatment of linear time-dependent systems was discussed briefly in Section 6.3.1, and Tong (1974a, 1974b) discusses the problem of controlling time-dependent linear systems.

Appendix

A.1. *The sunspot series (1700–1979)*

5.0	11.0	16.0	23.0	36.0	58.0	29.0	20.0	10.0	8.0
3.0	0.0	0.0	2.0	11.0	27.0	47.0	63.0	60.0	39.0
28.0	26.0	22.0	11.0	21.0	40.0	78.0	122.0	103.0	73.0
47.0	35.0	11.0	5.0	16.0	34.0	70.0	81.0	111.0	101.0
73.0	40.0	20.0	16.0	5.0	11.0	22.0	40.0	60.0	80.9
83.4	47.7	47.8	30.7	12.2	9.6	10.2	32.4	47.6	54.0
62.9	85.9	61.2	45.1	36.4	20.9	11.4	37.8	69.8	106.1
100.8	81.6	66.5	34.8	30.6	7.0	19.8	92.5	154.4	125.9
84.8	68.1	38.5	22.8	10.2	24.1	82.9	132.0	130.9	118.1
89.9	66.6	60.0	46.9	41.0	21.3	16.0	6.4	4.1	6.8
14.5	34.0	45.0	43.1	47.5	42.2	28.1	10.1	8.1	2.5
0.0	1.4	5.0	12.2	13.9	35.4	45.8	41.1	30.1	23.9
15.6	6.6	4.0	1.8	8.5	16.6	36.3	49.6	64.2	67.0
70.9	47.8	27.5	8.5	13.2	56.9	121.5	138.3	103.2	85.7
64.6	36.7	24.2	10.7	15.0	40.1	61.5	98.5	124.7	96.3
66.6	64.5	54.1	39.0	20.6	6.7	4.3	22.7	54.8	93.8
95.8	77.2	59.1	44.0	47.0	30.5	16.3	7.3	37.6	74.0
139.0	111.2	101.6	66.2	44.7	17.0	11.3	12.4	3.4	6.0
32.3	54.3	59.7	63.7	63.5	52.2	25.4	13.1	6.8	6.3
7.1	35.6	73.0	85.1	78.0	64.0	41.8	26.2	26.7	12.1
9.5	2.7	5.0	24.4	42.0	63.5	53.8	62.0	48.5	43.9
18.6	5.7	3.6	1.4	9.6	47.4	57.1	103.9	80.6	63.6
37.6	26.1	14.2	5.8	16.7	44.3	63.9	69.0	77.8	64.9
35.7	21.2	11.1	5.7	8.7	36.1	79.7	114.4	109.6	88.8
67.8	47.5	30.6	16.3	9.6	33.2	92.6	151.6	136.3	134.7
83.9	69.4	31.5	13.9	4.4	38.0	141.7	190.2	184.8	159.0
112.3	53.9	37.5	27.9	10.2	15.1	47.0	93.8	105.9	105.5
104.5	66.6	68.9	38.0	34.5	15.5	12.6	27.5	92.5	155.4

Note: The first row gives the annual sunspot numbers for the years 1700–1709, the second row gives the years 1710–1719, and so on.

A.2. *The Canadian lynx series (1821–1934)*

269	321	585	871	1475	2821	3928	5943	4950	2577	523
98	184	279	409	2285	2685	3409	1824	409	151	45
68	213	546	1033	2129	2536	957	361	377	225	360
731	1638	2725	2871	2119	684	299	236	245	552	1623
3311	6721	4254	687	255	473	358	784	1594	1676	2251
1426	756	299	201	229	469	736	2042	2811	4431	2511
389	73	39	49	59	188	377	1292	4031	3495	587
105	153	387	758	1307	3465	6991	6313	3794	1836	345
382	808	1388	2713	3800	3091	2985	3790	674	81	80
108	229	399	1132	2432	3574	2935	1537	529	485	662
1000	1590	2657	3396							

Note: The first row gives the data for the years 1821–1831, the second row gives the years 1832–1842, and so on.

Davis, R. C. (1952). On the theory of prediction of non-stationary processes. *J. Appl. Phys.*, **23**, 1047–1053.

De Schutter-Herteleer, A. and Melard, G. (1986). On time-dependent spectral concepts. Research report, Institut de Statistique and Centre d'Economie Mathématique et d'Econométrie, Université Libre de Bruxelles.

Fujita, T. and Shibara, H. (1978). On a model of earthquake ground motions for response analysis and some examples of analysis through experiment. *Conf. on Engineering Design for Earthquake Environments*, Vol. 12, pp. 139–148. I. Mech. Eng., London.

Gabr, M. M. and Subba Rao, T. (1981). The estimation and prediction of subset bilinear time series models with applications. *J. Time Series Anal.*, **2**, 153–171.

Giri, N. (1965). On the complex analogues of T^2 and R^2 tests. *Ann. Math. Statist.*, **36**, 664–670.

Granger, C. W. J. and Andersen, A. P. (1978). *An Introduction to Bilinear Time Series Models.* Vandenhoeck and Ruprecht, Göttingen.

Granger, C. W. J. and Hatanaka, M. (1964). *Spectral Analysis of Economic Time Series.* Princeton University Press, New Jersey.

Grenander, U. and Rosenblatt, M. (1957). *Statistical Analysis of Stationary Time Series.* Wiley, New York.

Haggan, V., Heravi, S. M. and Priestley, M. B. (1984). A study of the application of state-dependent models in non-linear time series analysis. *J. Time Series Anal.*, **5**, 69–102.

Haggan, V. and Ozaki, T. (1981). Modelling non-linear random vibrations using an amplitude-dependent autoregressive time series model. *Biometrika*, **68**, 189–196.

Hammond, J. K. (1968). On the response of single and multi degree-of-freedom systems to non-stationary random excitations. *J. Sound Vibr.*, **7**, 393.

Hammond, J. K. (1983). Evolutionary spectra in random vibrations. *J. Roy. Statist. Soc. Ser. B*, **35**, 167–188.

Hannan, E. J. (1967). Measurement of a wandering signal and noise. *J. Appl. Prob.*, **4**, 90–102.

Hannan, E. J. (1970). *Multiple Time Series.* Wiley, New York.

Harrison, P. J. and Stevens, C. F. (1976). Bayesian forecasting. *J. Roy. Statist. Soc. Ser. B*, **38**, 205–248.

Hasselman, K., Munk, W. and Macdonald, G. (1963). Bispectrum of ocean waves. In *Time Series Analysis* (Ed. M. Rosenblatt), pp. 125–139. Wiley, New York.

Hasofer, A. M. and Petocz, P. (1978). The envelope of an oscillatory process and its upcrossings. *Adv. Appl. Prob.*, **10**, 711–716.

Helland, K. N., Lii, K. S. and Rosenblatt, M. (1979). Bispectra and energy transfer in grid-generated turbulence. In *Developments in Statistics* (Ed. P. R. Krishaiha), Vol. 2, pp. 125–155. Academic Press, New York.

Hinnich, M. (1982). Testing for Gaussianity and linearity of a stationary time series. *J. Time Series Anal.*, **3**, 169–176.

Hitchings, D. and Beresford, P. J. (1978). A comparison of numerical methods for the aseismic design of mechanical systems. *Conf. Engineering Design for Earthquake Environment*, Vol. 12, pp. 129–138. I. Mech. Eng. London.

Housholder, A. S. (1964). *The Theory of Matrices in Numerical Analysis.* Blaisdell, New York.

Jenkins, G. M. (1961). General considerations in the estimation of spectra. *Technometrics*, **3**, 133–166.

Jenkins, G. M. and Priestley, M. B. (1957). The spectral analysis of time series. *J. Roy. Statist. Soc. Ser. B*, **19**, 1–12.

Jenkins, G. M. and Watts, D. G. (1968). *Spectral Analysis and its Applications.* Holden-Day, San Francisco.

Jones, D. A. (1978). Non-linear autoregressive processes. *Proc. Roy. Soc. London Ser. A*, **360**, 71-95.

Kalman, R. E. (1960). A new approach to linear filtering and prediction problems. *Trans. ASME J. Basic. Engng. Ser. D*, **82**, 35-45.

Kalman, R. E. (1963). Mathematical descriptions of linear dynamical systems. *J. SIAM Control Ser. A*, **1**, 152-192.

Kendall, M. G. and Stuart, A. (1966). *The Advanced Theory of Statistics*, 3 Vols. Griffin, London.

Khatri, C. G. (1965). Classical statistical analysis based on a certain multivariate complex Wishart distribution. *Ann. Math. Statist.*, **36**, 98-107.

Kiehm, J.-L. and Melard, G. (1981). Evolutive cospectral analysis of time-dependent ARMA processes. In *Time Series Analysis* (Eds O. D. Anderson and M. R. Perryman), pp. 227-236. North-Holland, Amsterdam.

Klimko, L. A. and Nelson, P. I. (1978). On conditional least squares estimation for stochastic processes. *Ann. Statist.*, **6**, 629-642.

Kolmogorov, A. (1941). Interpolation and extrapolation von stationären Zufälligen Folgen. *Bull. Acad. Sci.* (*Nauk*), *U.S.S.R.*, *Ser. Math.*, **5**, 3-14.

Koopmans, L. H. (1974). *The Spectral Analysis of Time Series.* Academic Press, New York and London.

Lawrance, A. J. and Lewis, P. A. W. (1985). Modelling and residual analysis of nonlinear autoregressive time series in exponential variables. *J. Roy. Statist. Soc. Ser. B*, **47**, 165-202.

Lii, K. S., Rosenblatt, M. and Van Atta, C. (1976). Bispectral measurements in turbulence. *J. Fluid Mech.*, **77**, 45-62.

Mark, W. D. (1970). Spectral analysis of the convolution and filtering of non-stationary stochastic processes. *J. Sound Vibr.*, **11**, 19.

Martin, W. (1981). Line tracking in nonstationary processes. *Signal Process.*, **3**, 147-155.

Melard, G. (1975). Processus purement indetérminables à paramètre discret: approaches fréquentielle et temporelle. Ph.D. thesis, Université Libre de Bruxelles.

Melard, G. (1978). Propriétés du spectra évolutif d'un processus non stationnaire. *Ann. Inst. Henri Poinconé, Sect. B*, **14**, 411-424.

Melard, G. (1985a). Analyse de données chronologiques. Séminaire de Mathématiques Supérieures—Sémaire Scientifique OTAN (NATO Advanced Study Institute) no. 89. Presses d l'Université de Montréal, Montréal.

Melard, G. (1985b). An example of the evolutionary spectrum theory. *J. Time Series Anal.*, **6**, 81-90.

Melard, G. and Wybouw, M. (1984). Analyse spectrale de modèles de fonction de transfert évolutifs. Paper presented at the Journées de Statistique, Montpellier La Grande-Motte, 21-24 May 1984.

Mohler, R. R. (1973). *Bilinear Control Processes.* Academic Press, New York and London.

Moran, P. A. P. (1953). The statistical analysis of the Canadian Lynx cycle: I—structure and prediction. *Austr. J. Zool.*, **1**, 163-173.

Morris, J. (1977). Forecasting the sunspot cycle. *J. Roy. Statist. Soc. Ser. A*, **140**, 437-447.

Ozaki, T. (1978). Non-linear models for non-linear random vibrations. Technical Report no. 92, Department of Mathematics, University of Manchester Institute of Science and Technology, UK.

Ozaki, T. (1981). Non-linear threshold autoregressive models for non-linear random vibrations. *J. Appl. Prob.*, **18**, 443–451.

Ozaki, T. (1982). The statistical analysis of perturbed limit cycle processes using nonlinear time series models. *J. Time Series Anal.*, **3**, 29–41.

Ozaki, T. (1985). Nonlinear time series models and dynamical systems. In *Handbook of Statistics* (Eds E. J. Hannan and P. R. Krishnaiah), Vol. 5. North-Holland, Amsterdam.

Page, C. H. (1952). Instantaneous power spectra. *J. Appl. Phys.*, **23**, 103–106.

Parthasarathy, K. R. (1961). On testing the mean of a discrete linear process. *Sankhyā*, **23**, 221–224.

Parzen, E. (1959). Statistical inference on time series by Hilbert space methods I. In *Time Series Analysis Papers.* Holden-Day, San Francisco.

Parzen, E. (1961). An approach to time series analysis. *Ann. Math. Statist.*, **32**, 951–989.

Priestley, M. B. (1965). Evolutionary spectra and non-stationary processes. *J. Roy. Statist. Soc. Ser. B*, **27**, 204–237.

Priestley, M. B. (1966). Design relations for non-stationary processes. *J. Roy. Statist. Soc. Ser. B*, **28**, 228–240.

Priestley, M. B. (1969). Control systems with time dependent parameters. *Bull. Inst. Int. Statist.*, **37**. (Paper presented at the 37th session of the ISI, London.)

Priestley, M. B. (1978a). System identification, Kalman filtering and stochastic control. In *Directions in Time Series* (Eds D. R. Brillinger and G. C. Tiao). I.M.S. Publication (1980).

Priestley, M. B. (1978b). Non-linear models in time series analysis. *The Statistician*, **27**, 159–176.

Priestley, M. B. (1980). State-dependent models: a general approach to non-linear time series analysis. *J. Time Series Anal.*, **1**, 47–71.

Priestley, M. B. (1981). *Spectral Analysis and Time Series*, 2 vols. Academic Press, London and New York.

Priestley, M. B. (1985). Discussion on "Modelling and residual analysis of nonlinear autoregressive time series in exponential variables". *J. Roy. Statist. Soc. Ser. B*, **47**, 188–189.

Priestley, M. B. and Chao, M. T. (1972). Non-parametric function filtering. *J. Roy. Statist. Soc. Ser. B*, **34**, 385–392.

Priestley, M. B. and Heravi, S. M. (1985). Identification of non-linear systems using general state-dependent models. *J. Appl. Prob.*, **23A**, 257–274.

Priestley, M. B. and Subba Rao, T. (1969). A test for stationarity of time series. *J. Roy. Statist. Soc. Ser. B*, **31**, 140–149.

Priestley, M. B. and Tong, H. (1973). On the analysis of bivariate non-stationary processes. *J. Roy. Statist. Soc. Ser. B*, **35**, 153–166.

Quenouille, M. H. (1958). The comparison of correlations in time series. *J. Roy. Statist. Soc. Ser. B*, **20**, 158–164.

Rao, A. G. and Shapiro, A. (1970). Adaptive smoothing using evolutionary spectra. *Management Sci.*, **17**, 208–218.

Rayleigh, Lord (1945). *The Theory of Sound.* Dover, American Edition.

Rosenbrock, H. H. (1970). *State-Space and Multivariate Systems.* Nelson, London.

Ruberti, A., Isidori, A. and d'Allessandro, P. (1972). *Theory of Bilinear Dynamical Systems.* Springer Verlag, Berlin.

Sen, P. K. (1965). Some non-parametric tests for non-dependent time-series. *J. Amer. Statist. Assoc.*, **60**, 134–147.

Shiryaev, A. N. (1960). Some problems in the spectral theory of higher order moments—I. *Theor. Prob. Appl.*, **5**, 265–284.

Stoker, J. J. (1950). *Nonlinear Vibrations.* Interscience, New York.

Subba Rao, T. (1968). A note on the asymptotic relative efficiency of Cox and Stuart's tests for testing trend in dispersion of *p*-dependent time series. *Biometrika*, **55**, 381–385.

Subba Rao, T. (1977). On the estimation of bilinear time series models. *Bull. Inst. Int. Statist.*, **41**. (Paper presented at the 41st session of the ISI, New Delhi.)

Subba Rao, T. (1979). On the theory of bilinear time series models—II. Technical Report no. 121, Department of Mathematics, University of Manchester Institute of Science and Technology, UK.

Subba Rao, T. (1981). On the theory of bilinear models. *J. Roy. Statist. Soc. Ser. B*, **43**, 244–255.

Subba Rao, T. (1981a). A cumulative sum test for detecting change in time series. *Int. J. Control*, **34**, 285–293.

Subba Rao, T. and Gabr, M. M. (1980). A test for linearity of stationary time series. *J. Time Series Anal.*, **1**, 145–158.

Subba Rao, T. and Gabr, M. M. (1984). *An Introduction to Bispectral Analysis and Bilinear Time Series Models.* Springer-Verlag, Berlin.

Subba Rao, T. and Nunes, A. M. D. (1985). Identification of non-linear (quadratic) systems using higher order spectra. Paper presented at the 7th IFAC Symposium on Identification and System Parameter Estimation, University of York, UK, July 1985.

Subba Rao, T. and Tong, H. (1972). A test for time-dependence of linear open-loop systems. *J. Roy. Statist. Soc. Ser. B*, **34**, 235–250.

Subba Rao, T. and Tong, H. (1973). On some tests for the time-dependence of a transfer function. *Biometrika*, **60**, 589–597.

Sussman, H. J. (1973). Minimal realisations of non-linear systems. In *Geometric Methods in Systems Theory* (Eds D. Q. Mayne and R. W. Brockett). Reidel, Dordrecht.

Sussman, H. J. (1976). Existence and uniqueness of minimal realisations of non-linear systems: I—initialised systems. In *Math. Syst. Theory* (Eds G. Morchesini and S. K. Mitter). Springer-Verlag, New York.

Tayfun, M. A., Yang, C. Y. and Hsiao, G. C. (1971). On non-stationary random wave spectra. *Int. Symp. on Stochastic Hydraulics*, Pittsburgh, May 1971.

Tick, L. J. (1961). The estimation of "transfer functions" of quadratic systems. *Technometrics*, **3**, 563–567.

Tjøstheim, D. (1976). Spectral generating operators for non-stationary processes. *Adv. Appl. Prob.*, **8**, 831–846.

Tong, H. (1974a). On time dependent linear transformations of non-stationary stochastic processes. *J. Appl. Prob.*, **11**, 53–62.

Tong, H. (1974b). Frequency domain approach to regulation of linear systems. *Automatica*, **10**, 533–538.

Tong, H. (1983). *Threshold Models in Non-linear Time Series Analysis.* Springer-Verlag, New York.

Tong, H. and Lim, K. S. (1980). Threshold autoregression, limit cycles and cyclical data. *J. Roy. Statist. Soc. Ser. B*, **42**, 245–292.

Tricomi, F. G. (1957). *Integral Equations.* Wiley, New York.

Trigg, D. W. and Leach, A. G. (1967). Exponential smoothing with an adaptive response rate. *Op. Res. Quart.*, **18**, 53–59.

Tuan, Pham-Dinh (1983). Bilinear Markovian representations and bilinear models. Technical Report no. 144, Department of Mathematics, University of Manchester Institute of Science and Technology, UK.

Tuan, Pham-Dinh and Lanh, Tat Tran (1981). On the first order bilinear time series model. *J. Appl. Prob.*, **18**, 617-627.

Tukey, J. W. (1959). An introduction to the measurement of spectra. In *Probability and Statistics* (Ed. V. Grenander), pp. 300-330. Wiley, New York.

Van der Pol, B. (1922). Forced oscillations in a system with non-linear resistance. *Phil. Mag.*, **3**, 65-80.

Volterra, V. (1959). *Theory of Functionals and of Integro-differential Equations.* Dover, New York.

Wahba, G. (1975). A canonical form for the problem of estimating smooth surfaces. Technical Report no. 420, Department of Statistics, University of Wisconsin.

Waldmeier, M. (1961). *The Sun-spot Activity in the Years 1610-1960.* Schulthess, Zürich.

Wilkinson, J. H. (1965). *The Algebraic Eigenvalue Problem.* Clarendon Press, Oxford.

Whittle, P. (1963). *Prediction and Regulation.* English Universities Press, London.

Whittle, P. (1965). Recursive relations for predictors of non-stationary processes. *J. Roy. Statist. Soc. Ser. B*, **27**, 523-532.

Wiener, N. (1949). *The Extrapolation, Interpolation and Smoothing of Stationary Time Series with Engineering Applications.* Wiley, New York.

Wiener, N. (1958). *Non-linear Problems in Random Theory.* MIT Press, Cambridge, Mass.

Woodward, R. H. and Goldsmith, P. L. (1964). *Cumulative Sum Techniques.* Oliver and Boyd, Edinburgh.

Yaglom, A. M. (1962). *An Introduction to the Theory of Stationary Random Functions.* Prentice-Hall, Englewood Cliffs, NJ.

Young, P. C. (1970). An instrumental variable method for real time identification of a noisy process. *Automatica*, **6**, 271-287.

Young, P. C. (1974). Recursive approaches to time series analysis. *Bull. Inst. Math. Appl.*, **10**, 209-224.

Young, P. C. (1975). Discussion on paper by Brown, Durbin, and Evans. *J. Roy. Statist. Soc. Ser. B*, **37**, 168-174.

Young, P. C. (1978). General theory of modelling for badly defined systems. In *Modelling, Identification and Control of Environmental Systems* (Ed. G. V. Vansteenkiste). North-Holland, Amsterdam.

Young, P. C. and Jakeman, A. (1979). Refined instrumental variable methods of recursive time series analysis, Part I: Single input-single output. *Int. J. Control*, **29**, 1-30.

Zadeh, L. A. (1953). Optimum non-linear filters. *J. Appl. Phys.*, **24**, 396-404.

Author Index

Subject Index

Department of Agricultural Economics and Rural Sociology
The Pennsylvania State University
University Park, Pennsylvania 16802